AutoSketch for Windows Release 2

AutoSketch for Windows Release 2

A. Yarwood

Autodesk.
Registered Developer

Longman

Longman
Edinburgh Gate, Harlow
Essex CM20 2JE, England
and Associated Companies throughout the world

First published 1996

British Library Cataloguing in Publication Data
A catalogue entry for this title is available from the British Library

ISBN 0-582-28577-1

Set by 24 in 10/13pt Melior
Produced by Longman Singapore Publishers (Pte) Ltd
Printed in Singapore

Contents

Preface

AutoSketch for Windows is an inexpensive computer-aided design (CAD) software package produced by Autodesk Limited, the publishers of AutoCAD. Although much cheaper than AutoCAD, AutoSketch does facilitiate construction of accurate and sophisticated two-dimensional (2D) drawings.

Unlike earlier versions of AutoSketch (versions 1, 2 and 3) which ran only under DOS (with the first two versions also available for the Archimedes) AutoSketch for Windows runs within the Windows operating environment and takes full advantage of the Windows graphical user interface (GUI) and its renowned performance.

AutoCAD is installed in more workstations than any other CAD software package, throughout the world. It has become the world leader in its field, outstripping the user bases of all other CAD systems. With this in mind, and given that AutoCAD Releases 12 and 13 are now also available in Windows versions, AutoSketch for Windows forms an economical introductory package for those wishing to develop expertise with AutoCAD. AutoSketch is suitable for all forms of 2D technical drawing – engineering, building, architecture and graphic design. The user can become proficient in using the software for constructing simple technical drawings within a relatively short period of time, although much practice is required to ensure that one can produce accurate and complex drawings of any nature.

I would recommend AutoSketch for Windows as a very suitable CAD software in educational establishments:

1. Wherever the requirements of the NCC Technology and Information Technology syllabuses demand a knowledge of CAD software.
2. For use as an introductory CAD software package where a knowledge of drawing with AutoCAD is required.
3. As a general introductory package for those wishing to learn how to construct technical drawings with any form of advanced CAD software.

The contents of this book form a basic course of work for those wishing to ensure they can construct drawings with CAD software. A description of the wide variety of dialogue and warning boxes seen on screen when the software is in operation, together with over seventy step-by-step examples of their use and a series of graded exercises for the reader, are included.

All the text and illustrations were produced with the aid of a Dart Computers 486DX2, equipped with 16 Mbytes RAM and a VESA-controlled video card, working in Windows 3.1 under MS-DOS 5.0. All line drawings were drawn with the aid of AutoSketch for Windows Release 2 and pasted directly into PageMaker 5.0 files through the Windows Clipboard prior to typesetting. The screen illustrations were produced with the aid of Hijaak 3.0 for Windows. The Hijaak illustrations were pasted as bitmaps directly into the PageMaker files.

Salisbury 1995 A. Yarwood

Acknowledgements

The author wishes to acknowledge with grateful thanks the help given to him by members of the staff of Autodesk Limited.

Trademarks

The following trademarks are registered in the US Patent and Trademark Office by Autodesk, Inc.:

Autodesk®, AutoCAD®, AutoSketch®, AutoSketch for Windows®, AutoCAD for Windows®, AutoCAD Release 12®, AutoCAD Release 13®, 3D Studio®.

IBM® is a registered trademark of the International Business Machines Corporation.

MS-DOS® is registered trademarks and Windows™ is a trademark of the Microsoft Corporation.

A. Yarwood is a Registered Developer with Autodesk Ltd.

Autodesk.

Registered Developer

CHAPTER 1

Finding your way around AutoSketch for Windows

The AutoSketch for Windows setup

1. AutoSketch for Windows will only run in Windows, which must be Windows 3.0 or higher.
2. The computer operating chip must be 386DX or higher.
3. AutoSketch will run faster if an 80387 math coprocessor chip is fitted – if the operating chip is 486DX or higher, the math co-processor will be an integral part of the chip.
4. The computer must have at least 2 Mbytes of RAM (on board memory).
5. The monitor must have a VGA (Video Graphic Array) screen or better.

Graphical User Interface

A prominent feature of Windows is its GUI (Graphical User Interface) system, in which icons (small picture-like shapes representing the features available to the operator) are a prominent feature. AutoSketch for Windows includes a large number of icons.

Windows and the mouse

Depending upon how Windows is configured, practically all operations in Windows are performed with the aid of a mouse. The mouse will most likely have two buttons – a left button and a right button. Some mice have three buttons. As the mouse is moved over the surface of the desk (or better over a mouse mat), a cursor in the shape of a floating arrow will move in sympathy on the computer screen.

Two actions of the mouse buttons feature prominently when working in Windows:

A *left-click* – the left-hand mouse button is pressed and immediately released.

A *right-click* – the right-hand button of the mouse is pressed and immediately released.

A *double-click* – the left-hand mouse button is pressed twice in quick succession.

When using AutoSketch for Windows, all drawing actions are carried out under the control of the mouse. Thus to draw a line:

1. Position the cursor, under the control of the mouse, over the icon representing line and *left-click.*
2. Position the cursor, under mouse control, at the point on screen where the line is to start and *left-click.*
3. Position the cursor, under mouse control, at the point on screen where the line is to end and *left-click* and the line is drawn.

Starting Windows

AutoSketch for Windows can only be started from Windows. Windows may be started in one of several ways:

1. The computer may be configured so that Windows comes into operation when the computer is switched on.
2. Windows is started by entering *win* at the **C:\>** MS-DOS prompt. In this case a *batch file* – **win.bat** – loads Windows.
3. If working in a school or college, a number of messages may appear on screen when the computer is started up telling the operator how to start up Windows.

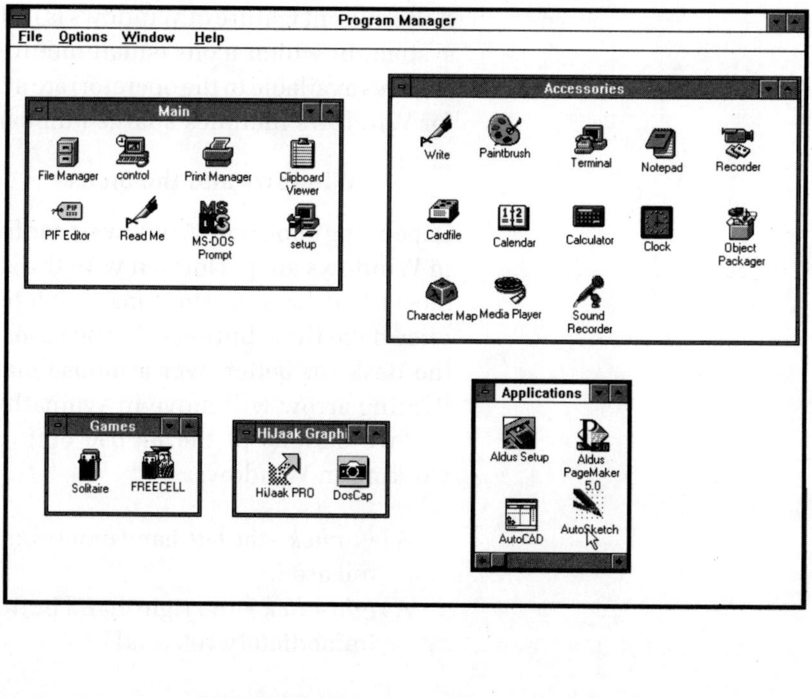

Fig. 1.1 The Windows **Program Manager** screen

The Windows Program Manager screen

When Windows is started up, the **Program Manager** screen appears (Fig. 1.1). This screen contains a number of icons representing the various software packages previously loaded into Windows and from which the operator can select the program he or she wishes to use. Fig. 1.1 shows a typical Windows **Program Manager** screen. The icon panel **Main** of the **Program Manager** shows, in icon form, the various programs within Windows itself – for configuring the way in which the Windows software operates. Other panels of icons in this screen show (in icon form) that four software packages (applications) are available for use:

1. The Aldus Setup program – the icon labelled **Aldus Setup**, together with the desktop application **PageMaker 5.0**.
2. A screen capture package called HiJaak Pro – the icon labelled **HiJaak PRO**.
3. AutoSketch for Windows – the icon labelled **AutoSketch**.
4. AutoCAD Release 12 – the icon labelled **AutoCAD**.

Note: Most Windows **Program Manager** screens will show a much larger number of icons than the example given above. Applications such as a word-processing program will normally be included. In the example given in Fig. 1.1, the icon for a word-processing application is hidden behind the **Main** panel.

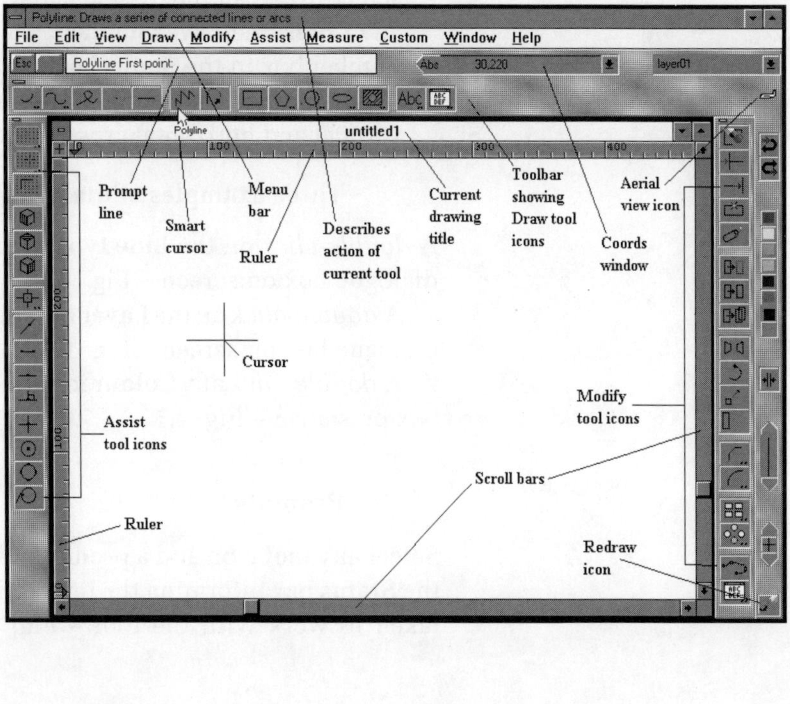

Fig. 1.2 The **AutoSketch for Windows** drawing screen

To start up AutoSketch for Windows, move the cursor over the **AutoSketch** icon *double-click* the left-hand button of the mouse and the AutoSketch drawing screen will appear.

The AutoSketch for Windows drawing screen

The AutoSketch for Windows drawing screen is shown in Fig. 1.2. In this example the commands showing are in icon form arranged in **Toolboxes**. A **Toolbox** is a group of **Tools** (commands) appearing as a collection of icons. In the example above, two **Toolboxes** – the **Draw** toolbox (horizontally) and the **Assist** and **Modify** toolboxes (vertically) are showing. Other toolboxes may appear upon start-up. The names of other toolboxes are shown in the menu bar – **File**, **Edit**, **View, Measure**, **Custom, Window**, and **Help**.

The following notes are helpful:

1. To bring a tool into action, move the cursor, under mouse control, over its icon and *left-click*.
2. Some icons show small double black rectangles at their bottom right corners. A *double-click* on any one of these icons will bring a dialogue box associated with the tool on screen. See later pages.
3. If configured to do so (see later), when the cursor is placed over a tool icon, a smart cursor will appear showing the function of that tool. One example – **Polyline** – is shown.
4. A *double-click* on any one of the following will also bring on to the screen a dialogue box associated with the feature shown by the icon: any **Color**; the **Layer number**; the **Line type** icon; the coordinate number box in the **Status bar**.
5. A *double-click* on the cross icon in the **Vertical status bar** brings vertical and horizontal crossing cursor lines on screen.

Three examples of dialogue boxes

A *double-click* on the **Line type** icon brings the **Line Type Settings** dialogue box on screen – Fig. 1.3.

A *double-click* on the **Layer number** icon brings the **Layer Settings** dialogue box on screen – Fig. 1.4.

A *double-click* any **Colour** icon brings the **Color Settings** dialogue box on screen – Fig. 1.5.

Prompts

Select any **tool** icon and a prompt will appear in the **Prompts** area of the **Status bar** informing the operator of the action which should be taken to work with the tool – Fig. 1.6. When several actions are

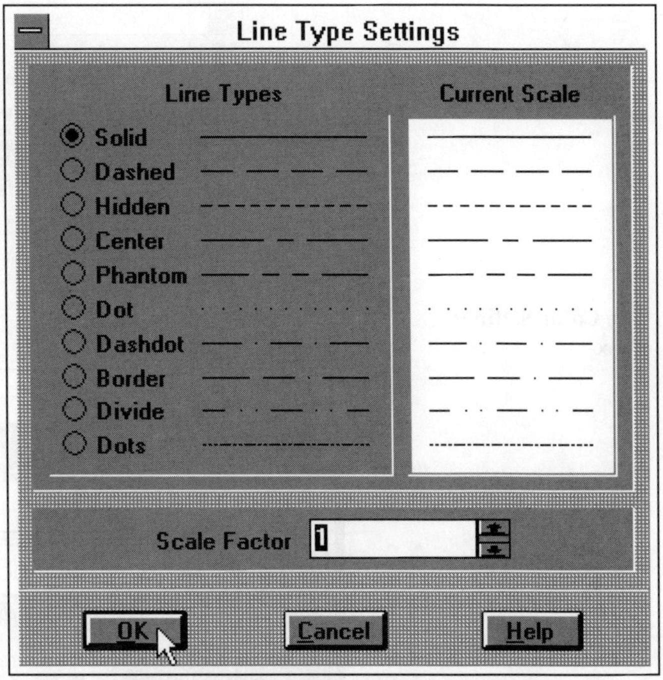

Fig. 1.3 The **Line Type Settings** dialogue box

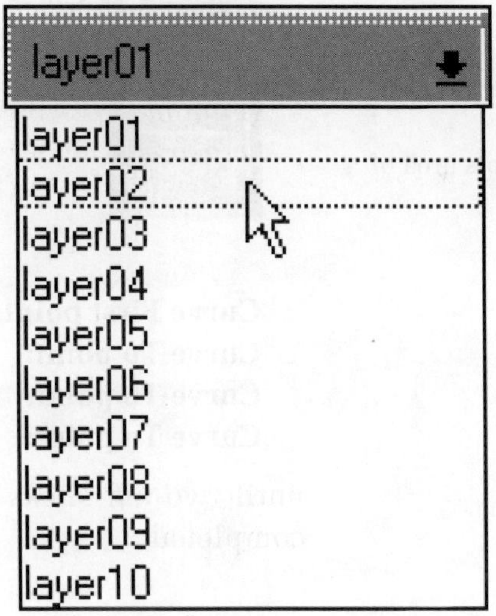

Fig. 1.4 The **Layer Settings** dialogue box

required to fully complete any tool operation, a series of prompts appear after each action has been taken. In the example given, the **Curve** tool has been selected. The following series of prompts appear in the **Prompts** area, one after the other:

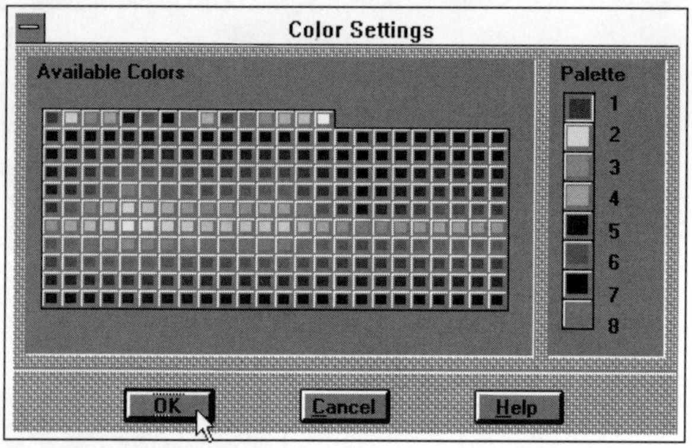

Fig. 1.5 The **Color Settings** dialogue box

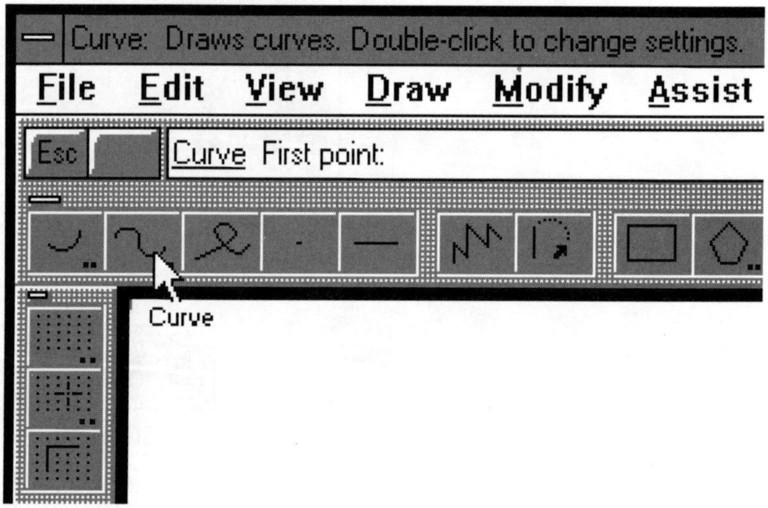

Fig. 1.6 The **Prompts** area of the **Status Bar**

Curve First point:
Curve To point:
Curve To point:
Curve To point:

until two *left-clicks* are made at the same point and the curve is completed.

Pull-down menus

A *left-click* on the button at the top left corner of the **Draw** toolbox – Fig. 1.7 – brings a new type of menu on screen – Fig. 1.8. A *left-click* on **Pulldown Menu** in this menu brings the **Draw** pull-down menu on screen – Fig. 1.9. Selections for tools can now be made from names of tools in this menu rather than from the icons in the **Draw** toolbox

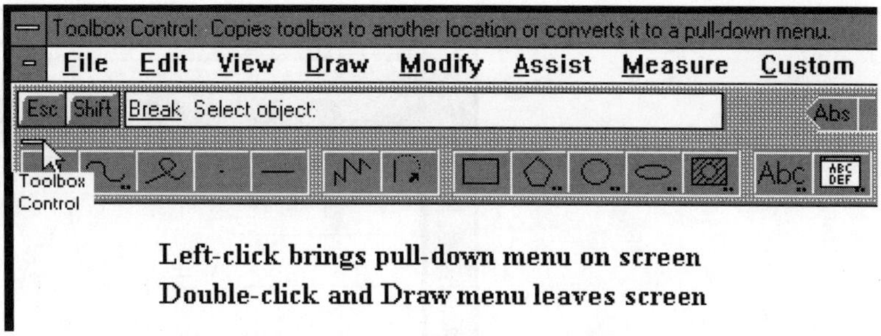

Fig. 1.7 The **Draw** toolbox

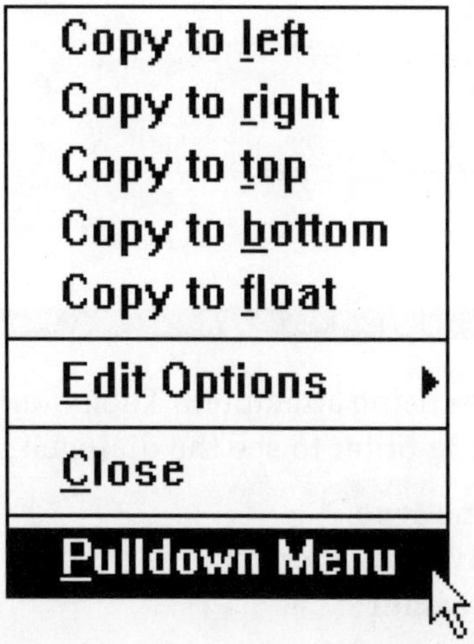

Fig. 1.8 Pull-down menu associated with any toolkit

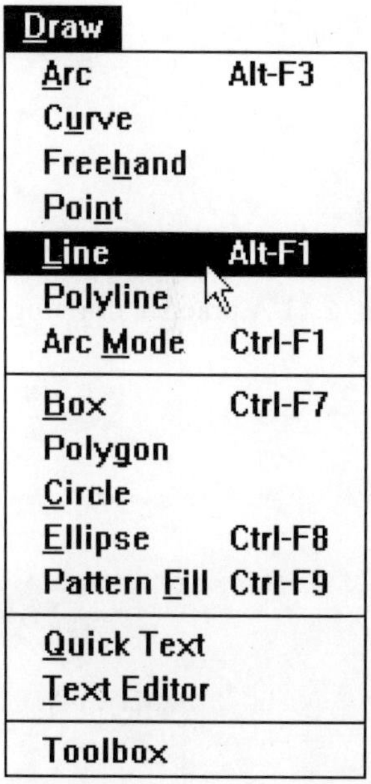

Fig. 1.9 The **Draw** pull-down menu

(Fig. 1.10). This same action applies to all other toolboxes. They can each be replaced by a pull-down menu if the operator prefers working with tool names rather than tool icons.

Practise drawing

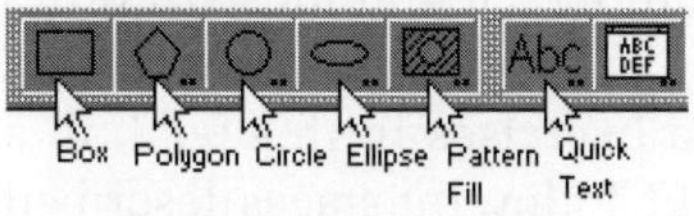

Fig. 1.10 Some of the tool icons from the **Draw** toolbox

Try each of the tools from the **Draw** toolbox in turn. Do not bother about making any particular shape or pattern; just place the cursor, under mouse control, on each tool in turn, *left-click* and draw any size or shape of arcs, curves, lines, polylines, boxes, circles and ellipses in order to 'get the feel' of using the tools – Fig. 1.11.

As you use each tool, note the prompts which appear in the **Prompts** area of the **Status bar** to guide you as to how the tool with which you are currently drawing should be used to produce the

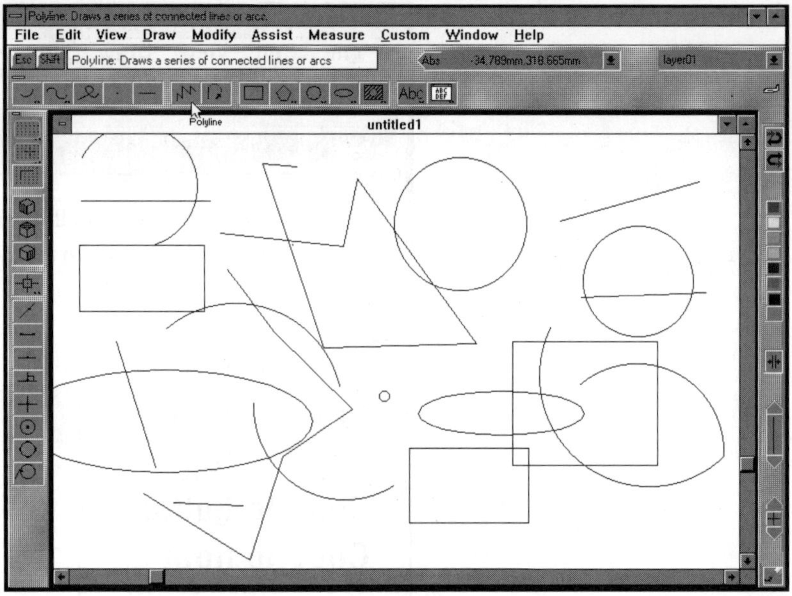

Fig. 1.11 A practice drawing

shapes being attempted. Then *double-click* on each of the following icons in order to see the dialogue boxes associated with them:

Line type
Layer number
A colour

Also *left-click* and *double-click* on the buttons in the top left corner of the **Draw** toolbox and the top left corner of the **Assist** toolbox (the toolbox currently at the left of the screen), in order to see the pull-down menus associated with these two toolboxes (Figs 1.8 and 1.9).

Setting up your own drawing sheet

The drawings and exercises in this book are worked in an AutoSketch graphics window configured as described below. When you have become reasonably proficient with using the software, you will find that there are many other AutoSketch graphics windows (templates) in which you can work. To set up the AutoSketch graphics window for the drawings and exercises in this book in a template which will be named **a3sheet.skt**, follow the stages described below.

Stage 1

1. Switch on the computer.
2. *Enter* win at the **C:\>** prompt.

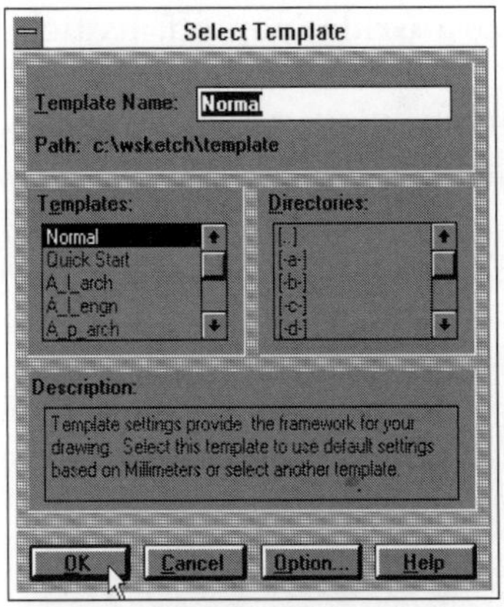

Fig. 1.12 The **Select Template** dialogue box

3. When the **Windows Program Manager** window appears, *double-left-click* on the **AutoSketch** application icon.
4. The **AutoSketch** graphics window appears with the **Select Template** dialogue box showing – Fig. 1.12.
5. *Double-left-click* on **Normal** in the list box of the dialogue box.

Stage 2

1. The dialogue box disappears and an AutoSketch graphics window appears (Fig. 1.13).

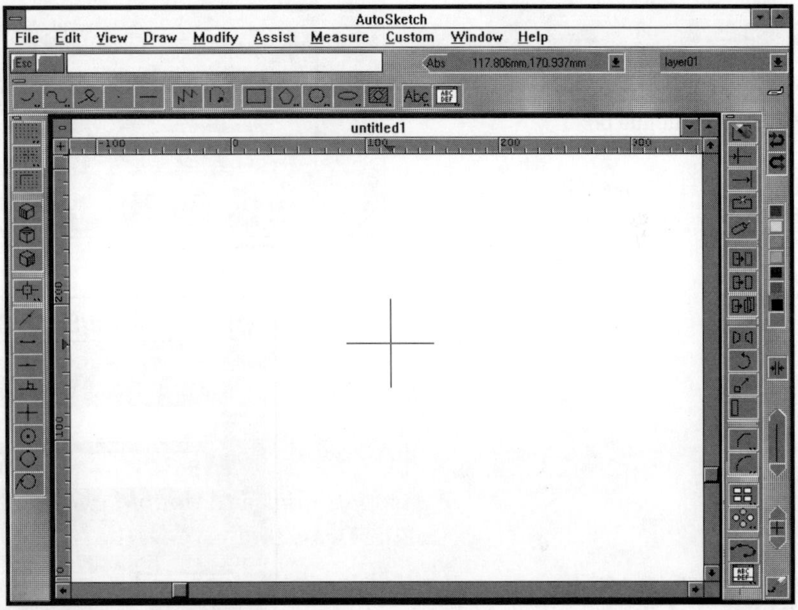

Fig. 1.13 The **Normal** AutoSketch graphics window

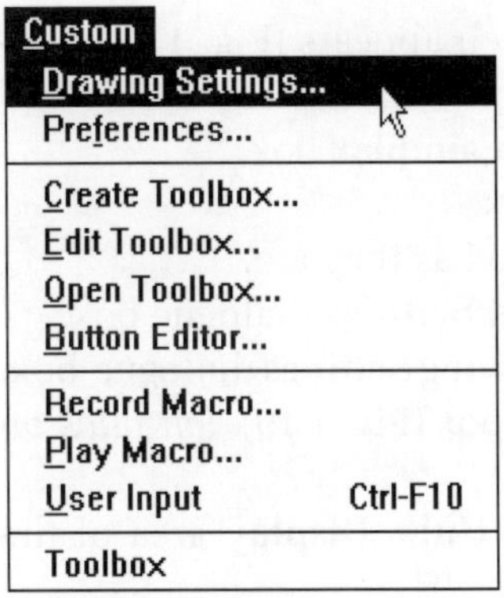

Fig. 1.14 Select **Drawing Settings...** from the **Custom** pull-down menu

2. *Left-click* on **Custom** in the menu bar.
3. In the **Custom** pull-down menu which appears *left-click* on **Drawing Settings...** (Fig. 1.14).
4. In the **Drawing Settings** dialogue box (Fig. 1.15) *left-click* on **Dimension....**
5. In the **Dimension Settings** dialogue box which appears (Fig. 1.16), *enter* 4 in both the **Text Size** and in the **Arrow Size** boxes. If the other settings are not as shown in Fig. 1.16, change them.
6. *Left-click* on the **OK button** of the **Dimension Settings** dialogue box.
7. *Double-left-click* on **Grid/Snap...** in the **Drawing Settings** dialogue box.
8. In the **Grid/Snap Settings** dialogue box which appears (Fig. 1.17), *left-click* in the **Snap** and **Grid** boxes to turn Grid and Snap on (a tick should appear in each box).

9. *Left-click* in the **Grid=Snap** box to turn **Grid=Snap** off. Its tick should disappear.
10. *Left-click* on the **OK** button of the **Grid/Snap Settings** dialogue box.
11. *Double-left-click* on **Text... in** the **Drawing Settings** dialogue box.

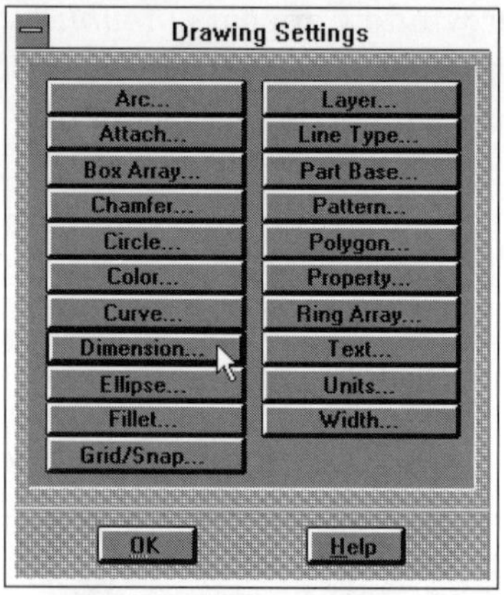

Fig. 1.15 The **Drawing Settings** dialogue box

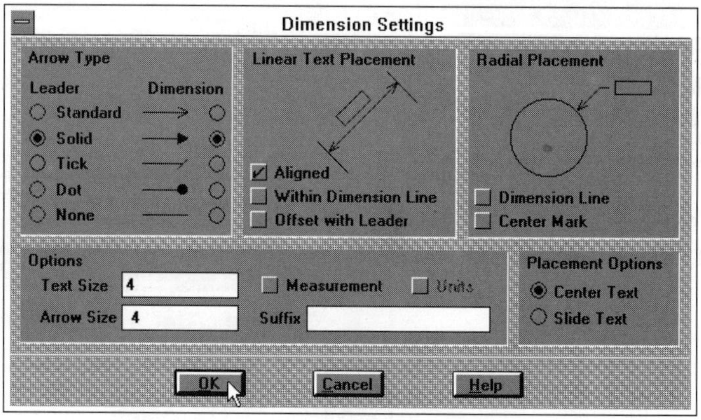

Fig. 1.16 The **Dimension Settngs** dialogue box

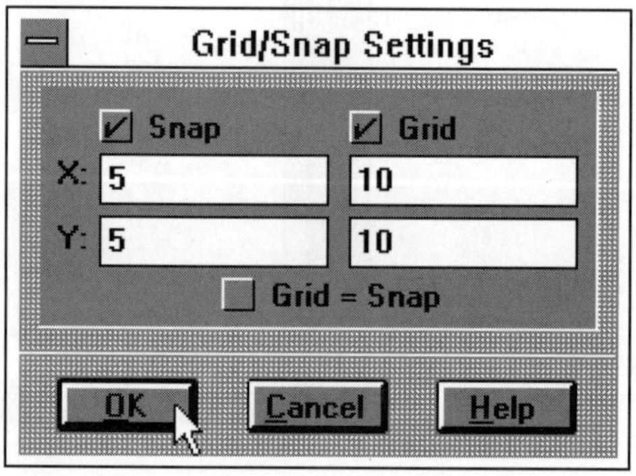

Fig. 1.17 The **Grid/Snap Settings** dialogue box

12. In the **Text Settings** dialogue box which appears (Fig. 1.18), *left-click* on **SHX Font** in the **Font Type** list box.
13. In the font style boxes *left-click* on the **simplex** box.
14. *Enter* 6 in the **Text Options Height** box.
15. Leave other settings in the dialogue box as they are.
16. *Left-click* on the **OK** button of the **Text Settings** dialogue box.
17. *Double-left-click on* **Units...** in the **Drawing Settings** dialogue box.
18. In the **Units of Measurement** dialogue box (Fig. 1.19), *left-click on* **Millimetres** in the **Unit Type** box.
19. *Left-click* on **1** in the list box in the **Units Display** area of the dialogue box.
20. Make sure there is no tick in the **Display Unit Type** box.

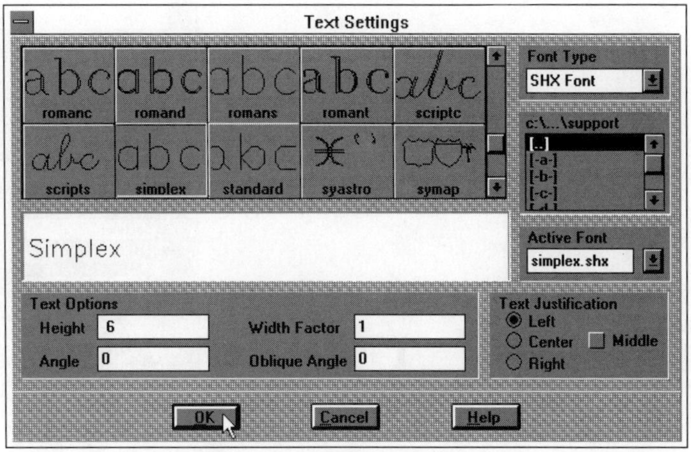

Fig. 1.18 The **Text Settings** dialogue box

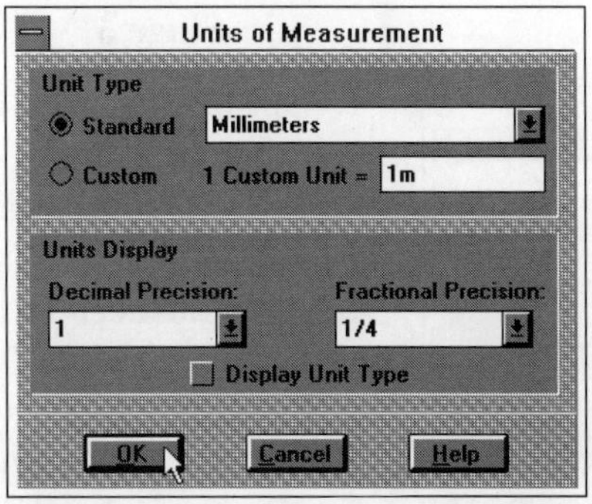

Fig. 1.19 The **Units of Measurement** dialogue box

Stage 3

1. *Left-click* on **View** in the menu bar.
2. In the **View** pull-down menu which appears (Fig. 1.20), *left-click* on **Pan**.
3. Move the cursor under mouse control until the figures in the **Coords** window show **0,0**.
4. Hold the left hand mouse button down while moving the cursor to the bottom left hand corner of the graphics window. Then release the mouse button. This pans the **Grid/Snap** system so that the coordinate X,Y = 0,0 is at the bottom left hand corner of the window.

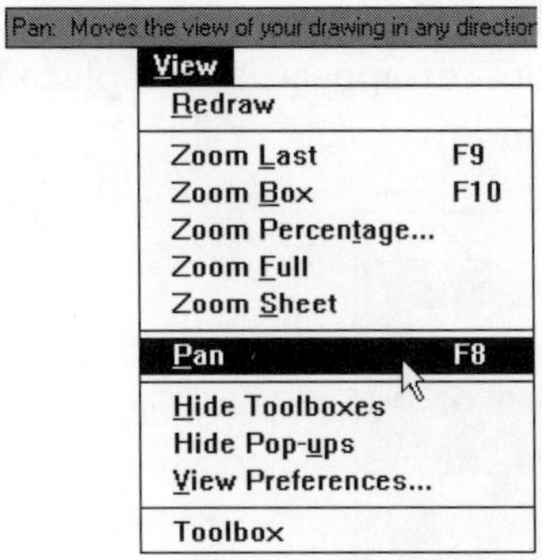

Fig. 1.20 Select **View** from the **View** pull-down menu

Stage 4

1. *Left-click* on **Preferences...** in the **Custom** pull-down menu.
2. In the **AutoSketch Preferences** dialogue box which appears (Fig. 1.21), make sure there is no tick in the **Show Rulers** box (*left-click* in the box if there is, to remove it).

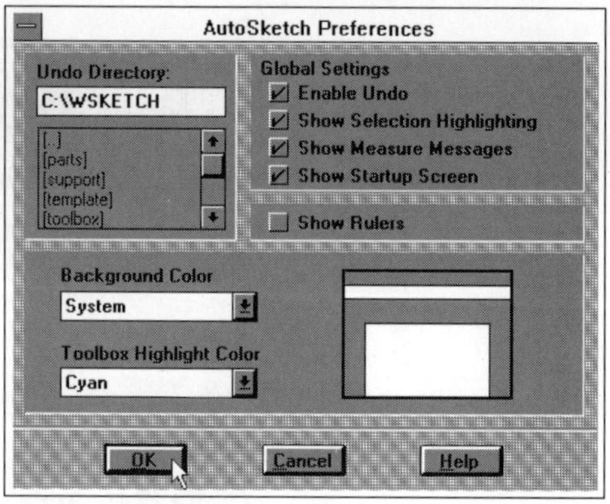

Fig. 1.21 The **AutoSketch Preferences** dialogue box

3. *Left-click* on the **OK** button of the **AutoSketch Preferences** dialogue box.

Stage 5

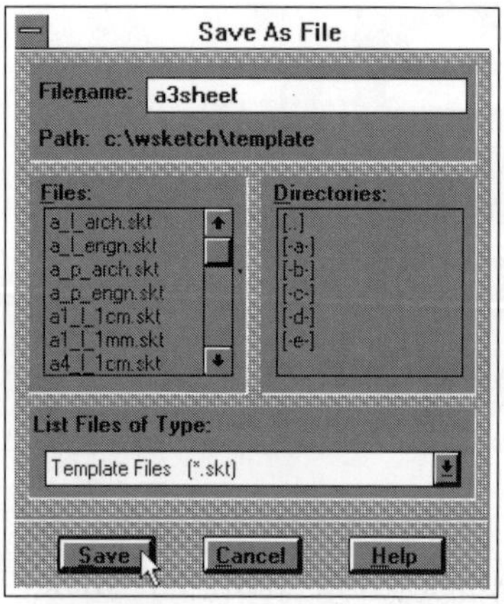

Fig. 1.22 The **Save As File** dialogue box

1. *Left-click* on **File** in the menu bar.
2. In the **File** pull-down menu which appears *left-click* on **Save As...**.
3. In the **Save As File** dialogue box which appears (Fig. 1.22), *left-click* on **Template Files [*.skt]** in the **List Files of Types** list box.
4. *Enter* **a3sheet** in the **Filename** box.
5. *Left-click* on the **OK** button of the dialogue box. The settings which have been made by following the stages outlined above have been saved to a **Template File** with a file name **a3sheet.skt** in the directory **c:\wsketch\template**.

Stage 6

1. *Double-click* in the button at the top left hand corner of the graphics window. The AutoSketch graphics window disappears to be replaced by the **Windows Program Manager** window.

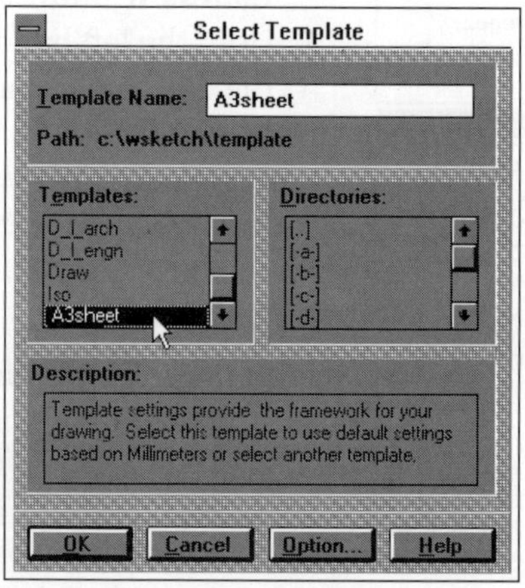

Fig. 1.23 The **Select Template** dialogue box

2. *Double-click* on the **AutoSketch** application icon.
3. The AutoSketch graphics window appears with the **Select Template** dialogue box showing (Fig. 1.23). This now includes your template file name **a3sheet**.
4. *Double-click* on the name **A3sheet** in the list box. The **a3sheet** template sheet appears in the AutoSketch graphics window (Fig. 1.24).
5. It is also best to save the template file to your own floppy disk in drive A: or B:.

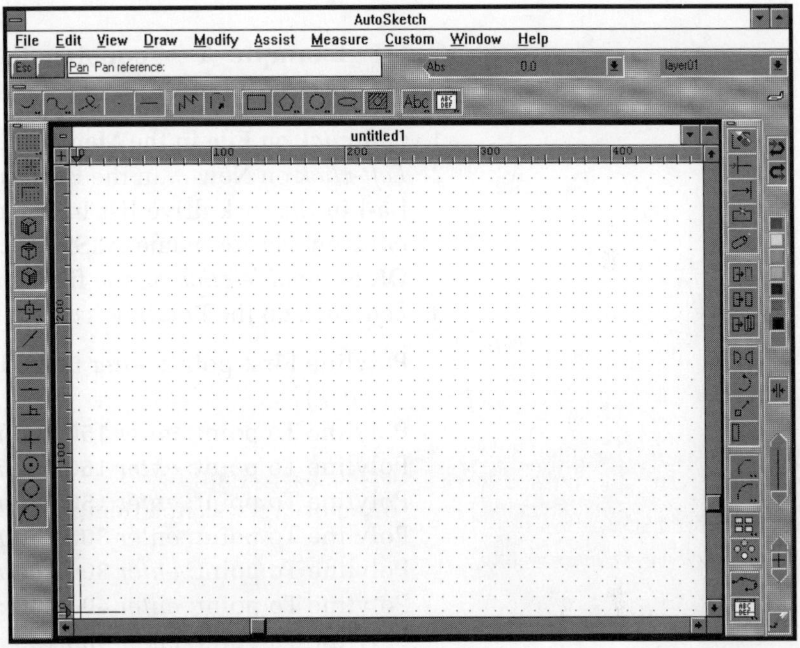

Fig. 1.24 The **A3sheet** template as it appears in the AutoSketch graphics window

CHAPTER 2

Constructing simple drawings

Example 1 – constructions using the absolute co-ordinate method

1. *Left-click* on **File** in the **Menu bar**.
2. *Left-click* on **New ...** in the **File** pull-down menu. *Double-click* on **[-a-]** in the disk drive list in **Select Template** dialogue box. *Left-click* on the file name **A3SHEET.SKT**, followed by *left-click* on **OK** or *double-click* on the file name in its box.
3. *Left-click* on the **Polyline** icon in the **Draw** toolbox.

Polyline First point: *enter* 50,200 *right-click* or *Return* (**not** *left-click*)
Polyline To point: *enter* 150,200 *right-click* or *Return*
Polyline To point: *enter* 150,200 *right-click* or *Return*
Polyline To point: *enter* 150,250 *right-click* or *Return*
Polyline To point: *enter* 300,250 *right-click* or *Return*
Polyline To point: *enter* 300,200 *right-click* or *Return*
Polyline To point: *enter* 400,200 *right-click* or *Return*
Polyline To point: *enter* 400,150 *right-click* or *Return*
Polyline To point: *enter* 300,150 *right-click* or *Return*
Polyline To point: *enter* 300,100 *right-click* or *Return*
Polyline To point: *enter* 150,100 *right-click* or *Return*
Polyline To point: *enter* 150,150 *right-click* or *Return*
Polyline To point: *enter* 50,150 *right-click* or *Return*
Polyline To point: *enter* 50,200 Polyline completed

4. *Left-click* on the **Circle** icon in the **Draw** toolbox.

Circle Centre point: *enter* 75,175 *right-click* or *Return* (**not** *left-click*)
Circle Point on circle: *enter* 75,185 *right-click* or *Return*
Circle Centre point: *enter* 375,175 *right-click* or *Return*
Circle Point on circle: *enter* 375,185 *right-click* or *Return*
Circle Centre point: *enter* 225,175 *right-click* or *Return*
Circle Point on circle: *enter* 225,215 *right-click* or *Return* – Fig. 2.1

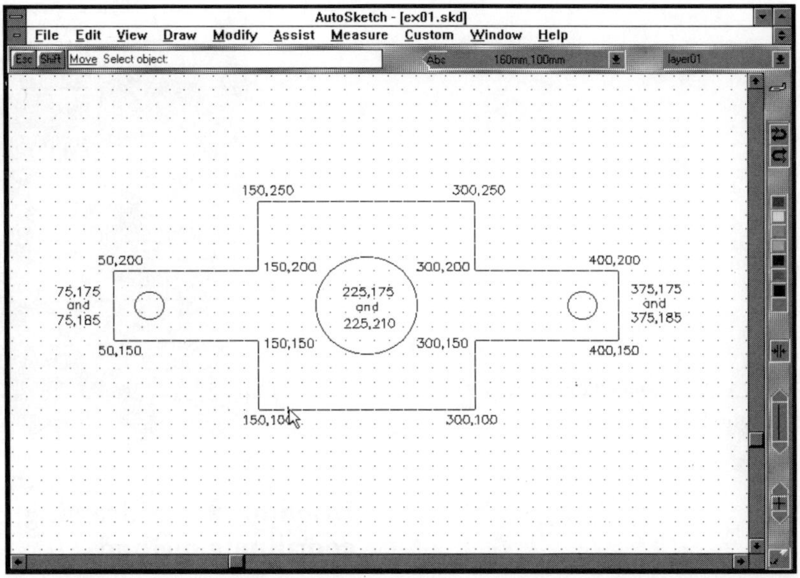

Fig. 2.1 Example 1

5. *Left-click* on **File** in the **Menu bar**. *Left-click* on **Save As ...** in the **File** pull-down menu – Fig. 2.2. *Left-click* on **[-a-]** in the **Directories** listbox in the **Save As File** dialogue box. *Enter* the file name **EX01** in the **Filename** box. *Left-click* on **OK** and the file is saved to disk in drive a:\ with the file name **EX01.SKD**.

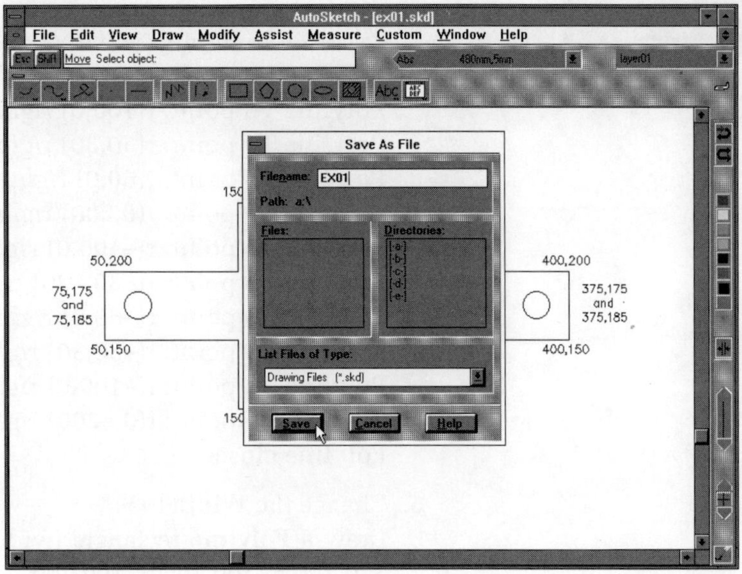

Fig. 2.2 The **Save As File** dialogue box

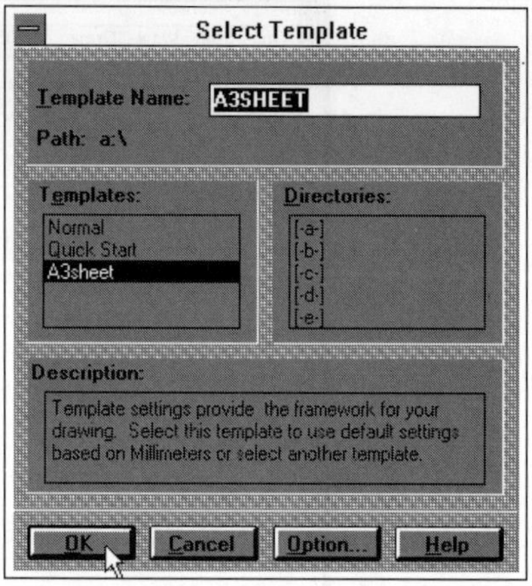

Fig. 2.3 The **Select Template** dialogue box

Example 2 – a drawing constructed by the relative coordinate method

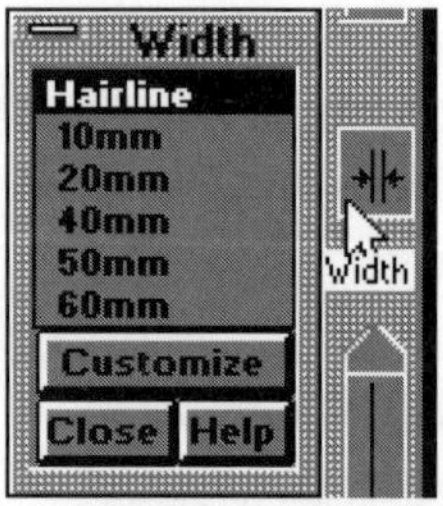

Fig. 2.4 The **Width** dialogue box

1. Load the drawing template **A3SHEET.SKT** – Fig. 2.3.
2. *Left-click* on the **Polyline** icon in the **Draw** toolbox.
3. *Left-click* on the **Width** icon, followed by *left-click* on **Hairline** (Fig 2.4). Then *enter* 3 in the **Width Settings** dialogue box that appears.
4. *Left-click* on the arrow marked **Abs** in the **Status bar** next to the coordinate numbers as many times as necessary to bring up **Rel** in the arrow.

Polyline First point: 50,50 *left-click* (or *Return*)
Polyline To point: r(100,0) *right-click*
Polyline To point: r(30,30) *right-click*
Polyline To point: r(60,0) *right-click*
Polyline To point: r(0,200) *right-click*
Polyline To point: r(–100,0) *right-click*
Polyline To point: r(–30,–30) *right-click*
Polyline To point: r(–60,0) *right-click*
Polyline To point: r(–30,30) *right-click*
Polyline To point: r(–100,0) *right-click*
Polyline To point: r(0,–200) *right-click*
Polyline closes

5. Change the **Width** to 6.
6. Draw a **Polyline** rectangle (width 6) from 100,200 to 320,200 to 320,100 to 100,100 to 100,200 – Fig 2.5.
7. **Save As ...** with file name **EX02.SKD.**

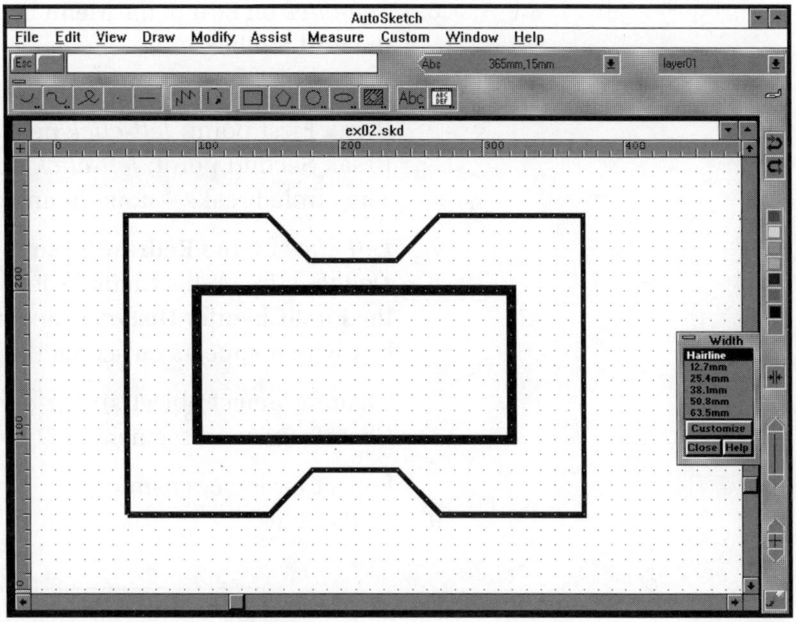

Fig. 2.5 Example 2

Example 3 – introducing tangents and chamfers

1. *Left-click* on **Draw** in the **Menu bar**. *Left-click* on the **Polyline** icon in the **Draw** toolbox.

   Polyline From point: 180,230 to 60,230 to 60,90 to 180,90

   Left-click twice to end the Polyline.
2. *Left-click* on **Assist** in the **Menu bar**. *Left-click* on the **Attach** icon in the **Assist** toolbox. *Left-click* on all **Attach** icons except **Tangent** to leave **Tangent** as the only active attach facility.
3. While still in the **Assist** toolbox, *left-click* on the **Snap** icon to turn **Snap** off.
4. *Left-click* on **Draw** in the **Menu bar**. *Left-click* on the **Line** icon in the **Draw** toolbox.

   Line from Point: 180,90 *right-click* (or *Return*)
   Line To point: *left-click* on lower part of circle. Tangential line drawn – tangent attach is on
   Line From point: 180,230 *right-click*
   Line To point: *left-click* on upper part of circle. Upper Tangentialline drawn

5. *Left-click* on **Assist** in the **Menu bar**. *Left-click* on all **Attach** icons except the **End Point** icon, to turn all attach points other than **End Point** off.

6. *Left-click* on **Edit** in the **Menu bar**. *Left-click* on the **Break** icon in the **Edit** toolbox.

   Break Select object: *left-click* on the circle
   Break First point: *left-click* near the end of the upper line
   Break Second point: *left-click* near the end of the lower line
   The circle breaks – note, in an anti-clockwise direction.

7. *Left-click* on the **Redraw** button.
8. *Double-click* on the **Chamfer** icon in the **Edit** toolbox. *Enter* 20 in the **First Chamfer Distance** box of the **Chamfer Settings** dialogue box which appears on screen (Fig. 2.6).

   Chamfer Select object(s): *left-click* on upper line
   Chamfer second segment: *left-click* on left-hand line

   Repeat for lower chamfer.
9. Add circles.

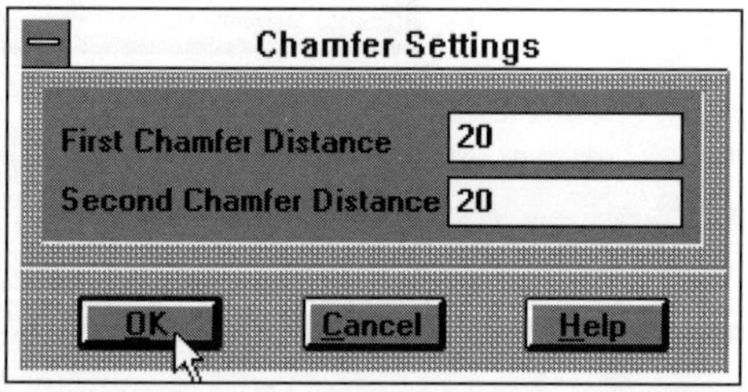

Fig 2.6 The **Chamfer Settings** dialogue box

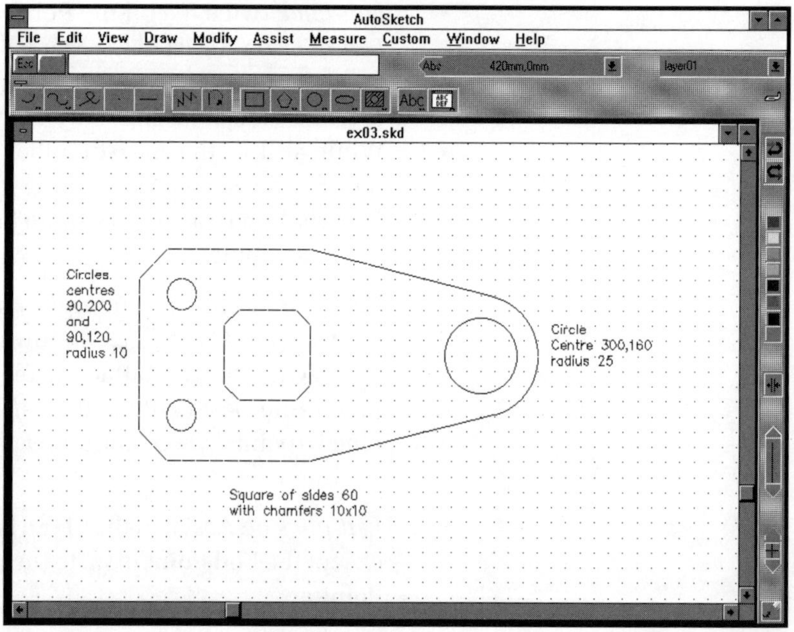

Fig. 2.7 Example 3

10. Add square and its chamfers – Fig. 2.7.
11. **Save As ...** with file name **EX03.SKD**.

Example 4 – introducing fillets

1. With **Polyline** draw the outline (Fig. 2.8).

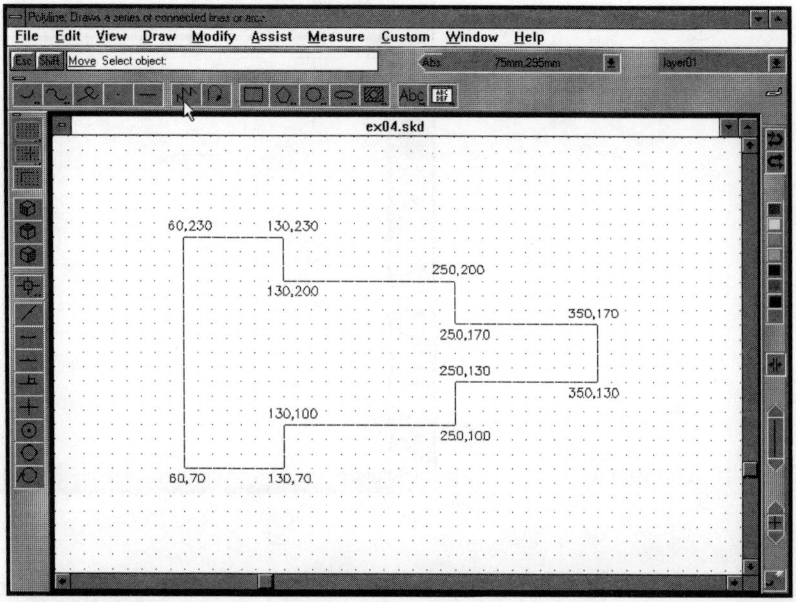

Fig. 2.8 First stage of Example 4

2. *Double-click* on the **Fillet** icon in the **Edit** Toolbox. *Enter* 15 in the **Fillet Radius** box of the **Fillet Settings** dialogue box (Fig. 2.9).
3. **Fillet Select object(s):** *left-click* on the upper line. **Fillet Select Second segment:** *left-click* on left-hand line. Add all R15 fillets in the same manner.
4. Re-set the **Fillet Radius** to 10 (Fig. 2.9). Add all R10 fillets.
5. Add all circles.
6. *Double-click* on the **Line type** icon in the **Vertical status bar**. *Left-click* on the small circle next to **Center** in the **Line Type Settings** dialogue box followed by *left-click* on **OK**.

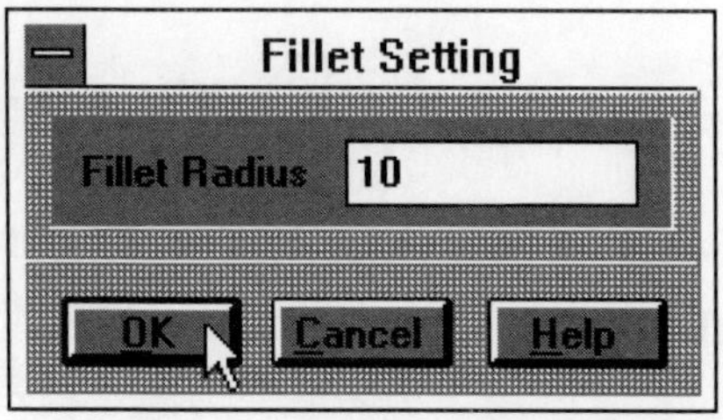

Fig. 2.9 The **Fillet Settings** dialogue box

7. *Left-click* on the colour **Red** icon in the **Vertical status bar**.
8. Add all centre lines through circles using the **Grid** and **Snap** points – Fig. 2.10.
9. **Save As ...** with file name **EX04.SKD**.

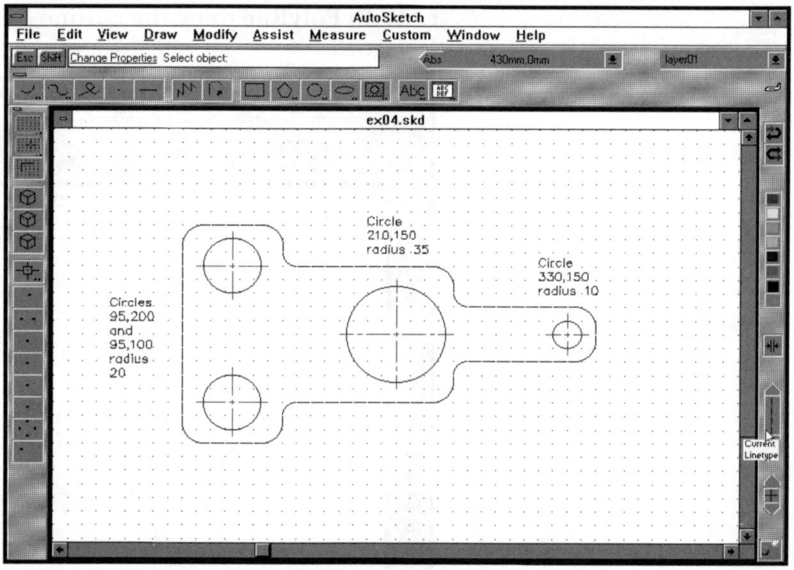

Fig. 2.10 Example 4

Example 5 – adding text to a drawing

1. **New ... A3SHEET.SKT**.
2. *Double-click* on the **Width** icon. *Enter* 20 in the **Width Settings** dialogue box (Fig. 2.11). *Left-click* on **Patterns...** button. *Left-click* on **circ** in the **Pattern Settings** dialogue box (Fig. 2.12). *Enter* 0.4 in the **Scale** box. *Left-click* on **OK**.

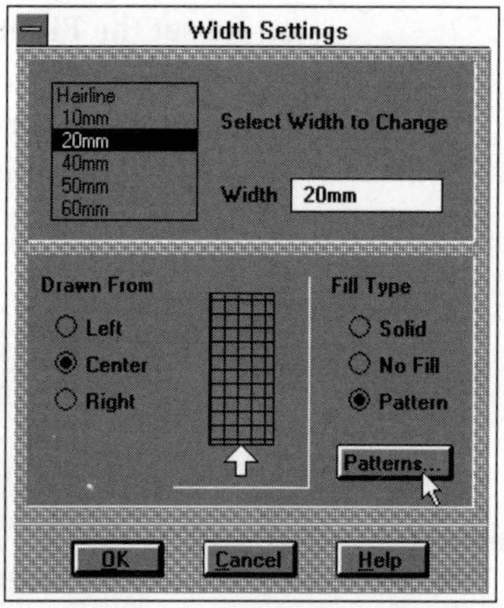

Fig. 2.11 The **Width Settings** dialogue box

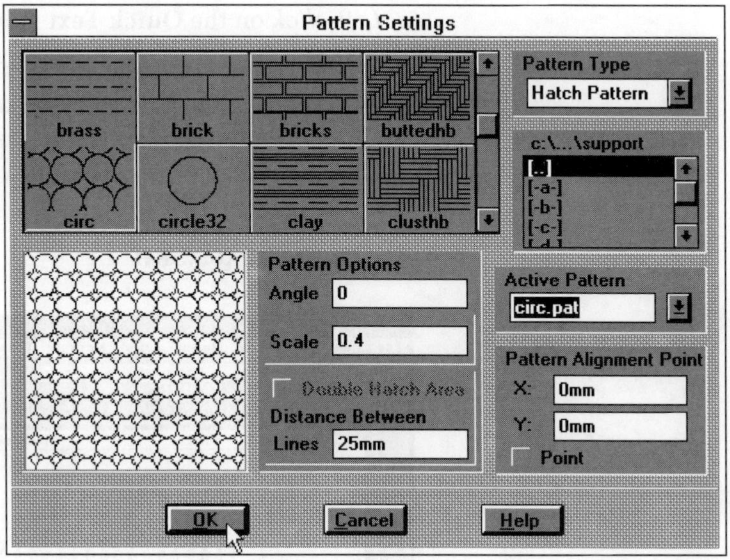

Fig. 2.12 The **Pattern Settings** dialogue box

3. *Left-click* on the **Polyline** icon in the **Draw** toolbox. Draw a polyline from 70,230 to 350,230 to 350,90 to 70,90 to 70,230.
4. *Double-click* on the **Quick Text** icon in the **Draw** toolbox. *Left-click* on the **Romant** box in the **Text Settings** dialogue box (Fig. 2.13). *Left-click* on the circle next to **Center** in the **Text Justification** box. *Enter* 20 in the Height box. *Left-click* on **OK**.

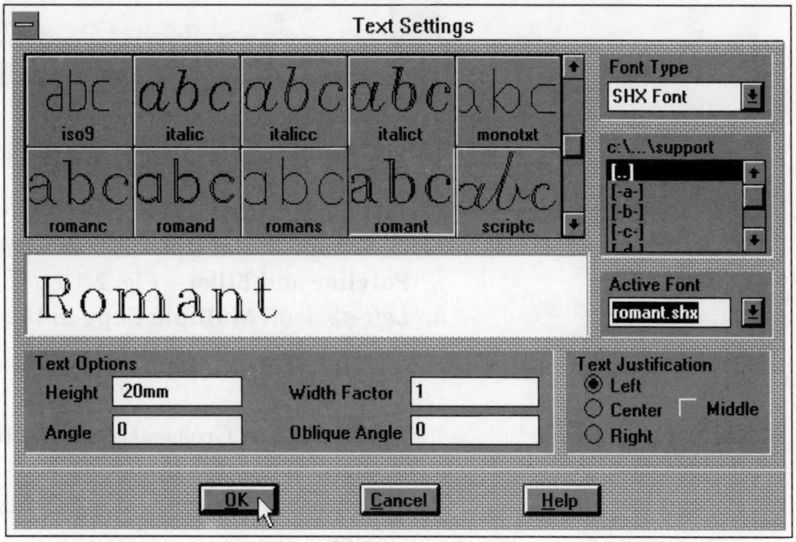

Fig. 2.13 The **Text Settings** dialogue box

5. *Left-click* on the **Quick Text** icon in the **Draw** toolbox.

   Quick Text Enter Point: 205,185 *Return* (note: **not** a mouse *click*)
   Quick Text Enter Text: AutoSketch *right-click*
   Quick Text Enter Text: for *right-click*
   Quick Text Enter Text: Windows *Return* (note: **not** a mouse *click*) – Fig. 2.14

6. **Save As ... EX05.SKD**

Fig. 2.14 Example 5

Example 6 – copying parts of drawings

1. **New ... A3SHEET.SKT**.
2. Construct the outline and the centrally place shape – use **Box**, **Polyline** and **Fillet** – Fig. 2.15.
3. *Left-click* on **Multiple Copy** in the **Edit** toolbox.

   Multiple Copy Crosses/window corner: *left-click* above and to left of shape
   Multiple Copy Crosses/window corner: *left-click* below and to the right of central shape
   Multiple Copy From point: *left-click* anywhere near (or inside) shape
   Multiple Copy To point: *left-click* to required new position
   Multiple Copy To point: *left-click* to required new position

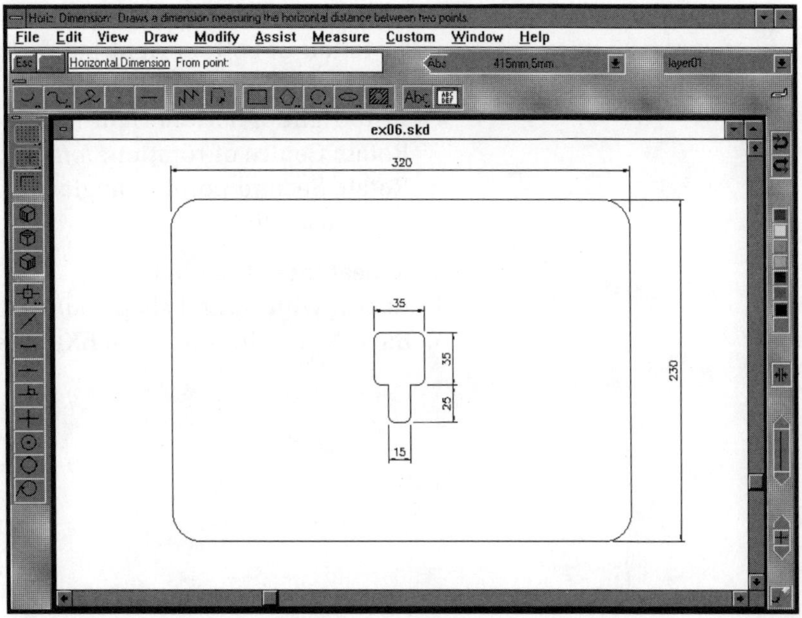

Fig. 2.15 First stage of Example 6

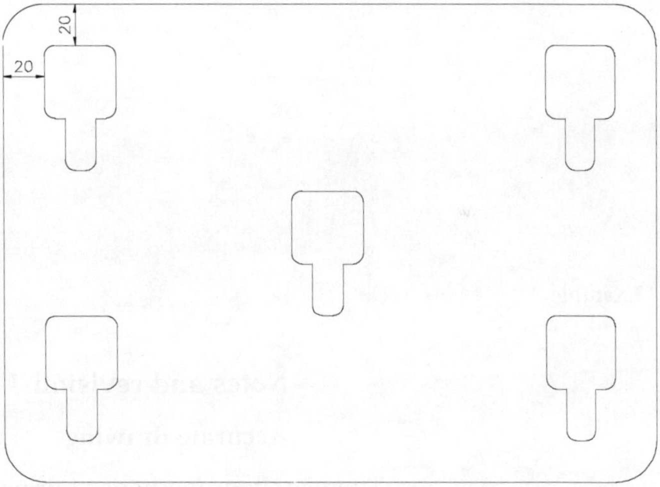

Fig. 2.16 Second stage of Example 6

Multiple Copy To point: *left-click* to required new position
Multiple Copy To point: *left-click* on **Multiple Copy** icon to end the copying – Fig. 2.16

4. **Save As ...** with file name **EX06.SKD**

Example 7 – rotating parts of drawings

1. **Open ... EX06.SKD**.
2. *Left-click* on the **Rotate** icon in the **Edit** toolbox.

Rotate Select object Crosses/window corner: *left-click* above and to left of bottom right-hand shape
Rotate Select object Crosses/window corner: *left-click* below and to right of bottom right-hand shape
Rotate Centre of rotation: *left-click* at centre of shape
Rotate Second point or angle (Angle 0): *enter* (180) *Return* (**not** a mouse *click*)

3. Repeat for bottom left shape.
4. Repeat with central shape but with Angle 90 – Fig. 2.17.
5. **Save As ...** with file name **EX07.SKD**

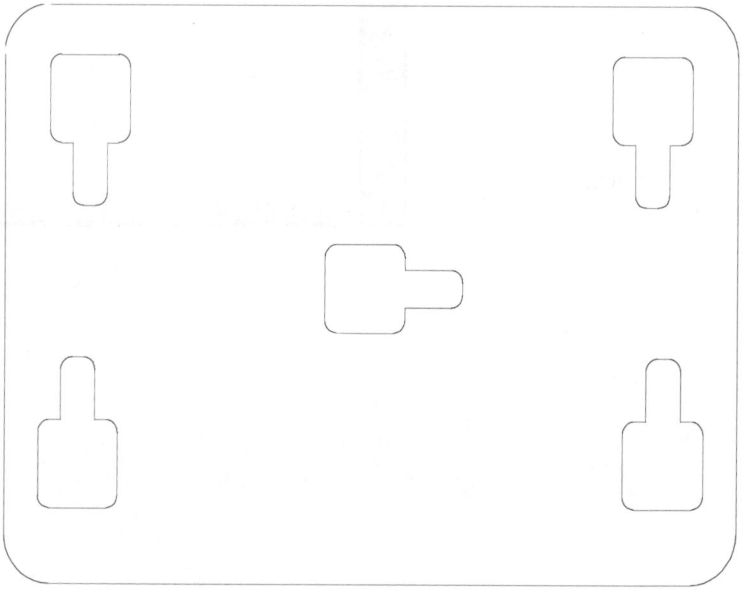

Fig. 2.17 Example 7

Notes and revision

Accurate drawing

There are four methods of drawing objects (e.g. Lines, Circles, Arcs etc.) in AutoSketch in order to ensure that accurate dimensions are obtained. These are:

1. The **Absolute** method of using *x,y* coordinates – **Example 1** (page 14). The pair of coordinate numbers with a comma between the *x* and *y* values are entered from the keyboard for each point on the object.
2. The **Relative** method – **Example 2** (page 15). The pair of *x,y* values are placed in brackets preceded by an r (for Relative) – e.g. r(100,60) will set the next point on the object at 100 units in the *x* direction

and 60 units in the *y* direction from the last point entered. The sign conventions are as follows:

Lengths to the right will be in +ve x units.
Lengths to the left will be in –ve x units.
Lengths vertically upwards will be in +ve y units.
Lengths vertically down will be in –ve *y* units.

For example, r(–100,–50) will be 100 units to the left and 50 units below the last point; r(50,100) will be 50 units to the right and 100 units above the last point.

3. With **Snap** (and preferably **Grid**) set, moving the cursor under mouse control until the required position of the cursor shows in the **Status bar**.
4. Not so far mentioned – the **Polar** method. The required next point is given as a length and an angle from the last position – e.g. p(100,45) is a point 100 units at an angle of 45 degrees from the last point. The p preceding the brackets is for Polar, the first number in the brackets is the unit length and the second the angle in degrees. Note that the polar degree measurements are from 0 to 360 degrees:

   0 degrees and 360 degrees due east (3:00 o'clock)
   90 degrees due north (12:00 o'clock)
   180 degrees due west (9:00 o'clock)
   270 degrees due south (6:00 o'clock)

   Note that when brackets ( ) enclose *x,y* values, if a bracket is missed, a warning message will appear on screen.

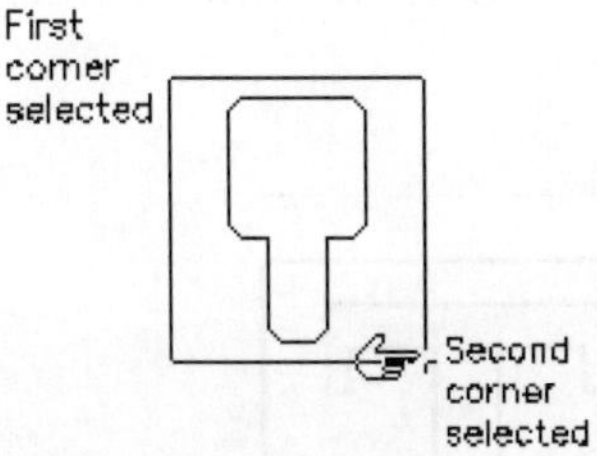

Fig. 2.18 Windowing a drawing top left to bottom right

The tools Move and Copy

In Example 6 (page 23) the tool **Multiple Copy** was used to copy part of a drawing. When the two tools **Copy** and **Move** are used, the same methods are involved as for **Multiple Copy**. Usually a number of lines, circles, etc., require moving or copying and they must be placed in a **window** to facilitate the action. The corner of the window is indicated by a pointing finger, with the lines of the **window** held and moved by the finger.

There are two types of window:

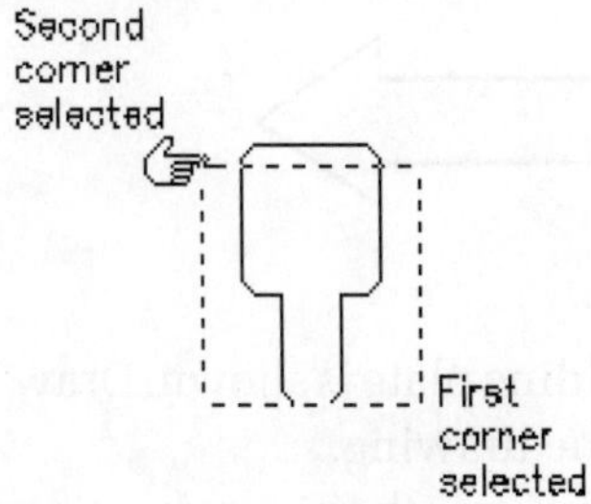

Fig. 2.19 Windowing a drawing bottom right to top left

1. Corners left to right: if the window corners are selected from left to right, the window outlines show as solid lines and only those objects **inside** the window will be selected – Fig. 2.18.
2. Corners right to left: if the window corners are selected from right to left, the window outlines show as broken lines and all objects **crossed** by the lines of the window will be selected – Fig. 2.19.

Revision exercises

The following four exercises are intended to be worked in with the screen configured as for the **A3SHEET.SKT** drawing already described. The settings are:

Grid 10
Snap 5
Limits 420,296
Units 0 digits after decimal point
Line Type Scale 1

Each exercise drawing includes the names of the tools which may be used to complete the answer drawing. Dimensions are included with the drawings only for the purpose of informing the reader of the sizes to which he or she should work. Do not attempt as yet to add dimensions to your answers. Practice in dimensioning drawings will come later in the book.

Exercises

1. Fig. 2.20. Two direction signs. Draw the two signs to the dimensions given.

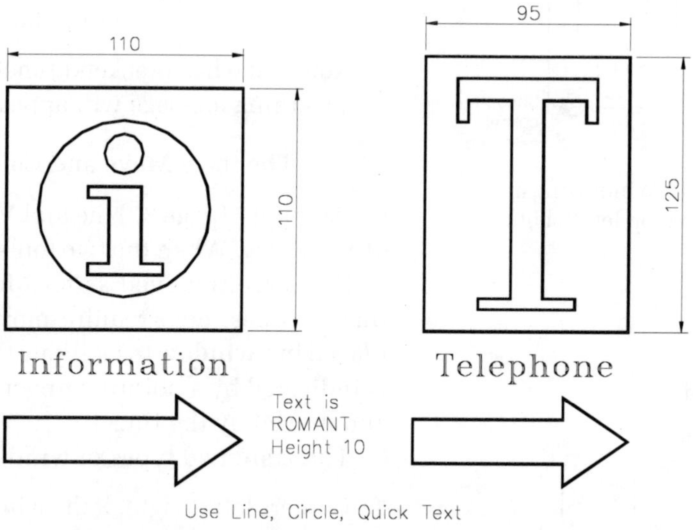

Fig. 2.20 Exercise 1

2. Fig. 2.21. An outline drawing of a steel holding plate is shown. Draw the plate to the dimensions given with the drawing.
3. Fig. 2.22. A front view of a house is shown. Copy the given drawing to the dimensions included with the drawing.
4. Fig. 2.23. The outline of a plastic sheet plate is given. Copy the drawing to the dimensions included with the drawing.

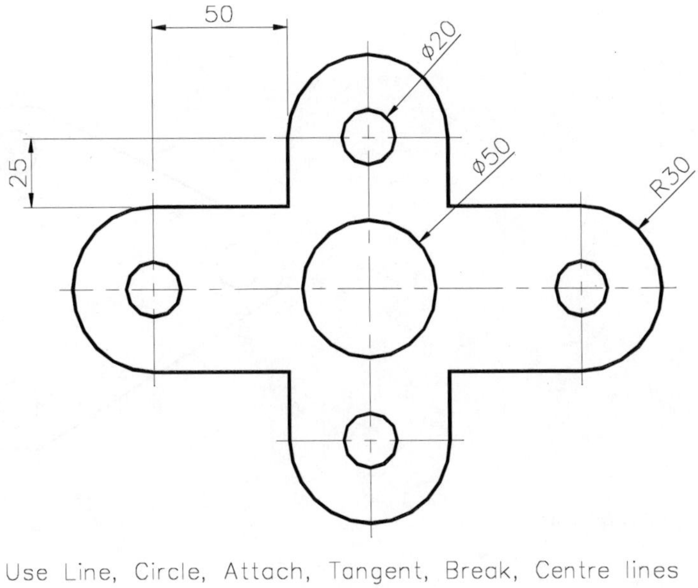

Fig. 2.21 Exercise 2

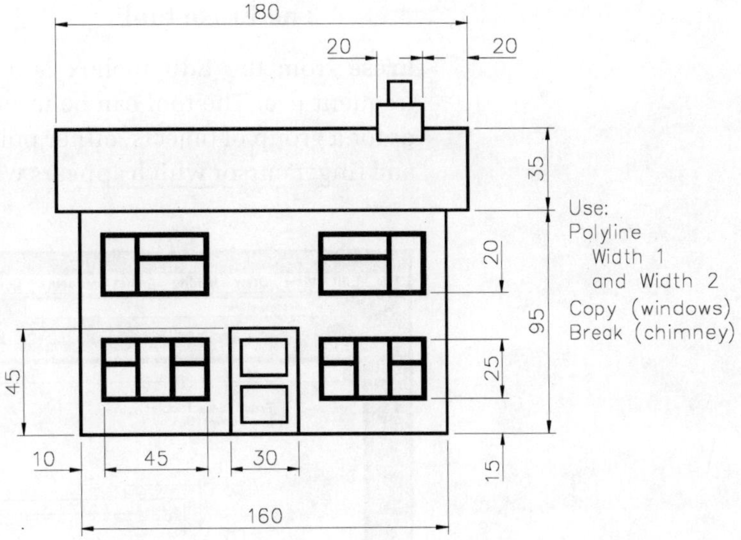

Fig. 2.22 Exercise 3

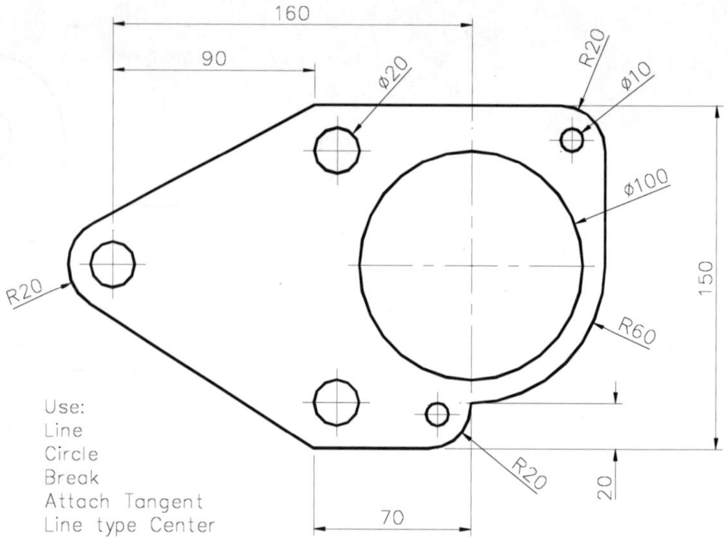

Fig. 2.23 Exercise 4

The Erase tool

Erase from the **Edit** toobox is a tool which will probably be in frequent use. The tool can be used for the erasure of a single object or for a group of objects, either pointing on the object with the hand and finger cursor which appears when the tool is selected (Fig. 2.24),

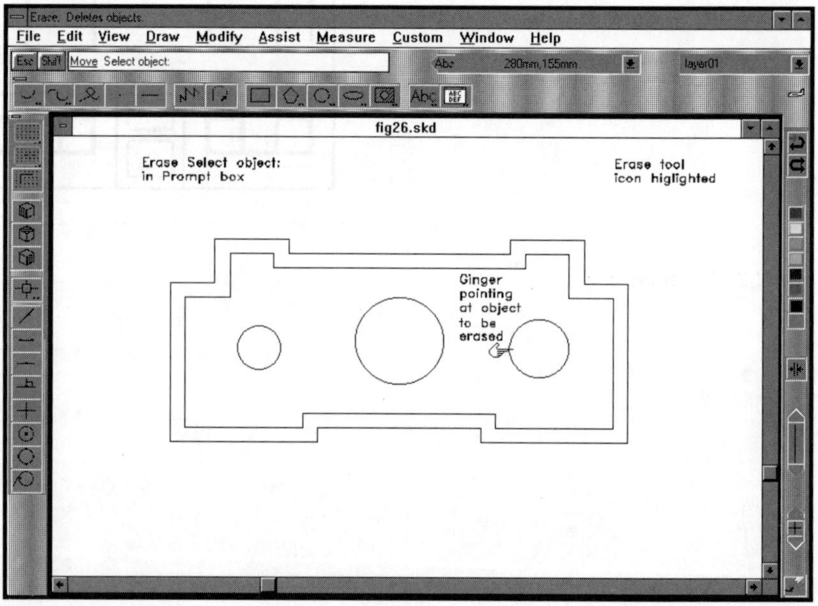

Fig. 2.24 Erasing an object

or using a window – either a left-to-right or a right-to-left window (as for **Move** and **Copy**). The procedure is:

Left-click on **Erase** in the **Edit menu**.
Erase Select object: *left-click* on the object. The object is removed from the screen.
Erase Select object:

Alternatively:

Erases elect object: *left-click* above and to the left (or right and below for a crossing window).
Erase Crosses/window corner: *left-click* below and to the left (or right and above for a crossing window).
Erase Select object: the objects within, or crossed by, the window are removed from the screen.
Erase Select object.

CHAPTER 3

Other Draw and Edit tools

In the following examples all coordinate points are found by moving the screen cursor under mouse control on to the given *x,y* coordinates.

Example 1 – Arc, Curve and Ellipse tools from the Draw toolbox

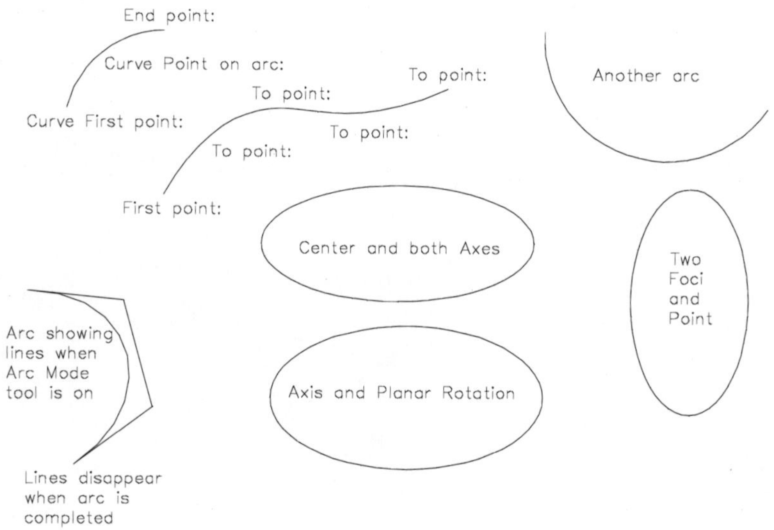

Fig. 3.1 Example 1

Fig. 3.1 shows arcs, curves and elipses drawn as follows:

1. **New ...** the file **A3SHEET.SKT**.
2. *Left-click* on the **Arc** icon in the **Draw** toolbox – Fig. 3.2.

   Arc First point: *left-click* at point *x,y* = 40,230
   Arc Point on arc: *left-click* at point *x,y* = 60,230
   Arc End point: *left-click* at point *x,y* = 90,270

   The arc now appears on the screen.
3. *Left-click* on the **Curve** icon in the **Draw** toolbox – Fig. 3.3.

Fig. 3.2 The **Arc** tool icon

Fig. 3.3 The **Curve** tool icon

Curve First point: *left-click* at point *x,y* = 90,105
Curve To point: *left-click* at point *x,y* = 110,220
Curve To point: *left-click* at point *x,y* = 150,240
Curve To point: *left-click* at point *x,y* = 190,220
Curve To point: *double-click* at point *x,y* = 230,240

The curve is now drawn on the screen. Note that *left-click* on the **Arc Mode** icon in the **Assist** toolbox before drawing a curve and lines between the selected points for the curve appear on the screen before the curve appears. These lines frame the curve.

Fig. 3.4 The **Ellipse** tool icon

4. *Double-click* on the **Ellipse** tool in the **Draw** toolbox – Fig. 3.4. *Left-click* on the circle next to **Center and Both Axes** in the **Ellipse Settings** dialogue box (Fig. 3.5), followed by *left-click* on **OK**.

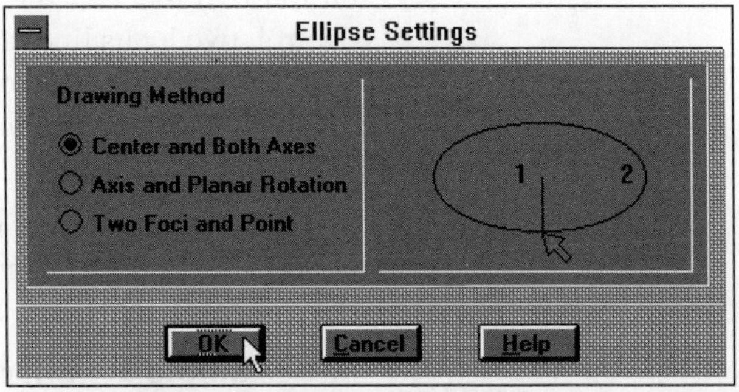

Fig. 3.5 The **Ellipse Setting** dialogue box

5. *Left-click* on the **Ellipse** icon in the **Draw** toolbox.

   Ellipse Center of ellipse: *left-click* on 210,160
   Ellipse Axis endpoint: *left-click* on 280,160
   Ellipse Other axis distance: *left-click* on 210,160

   The ellipse now appears on the screen.
6. *Double-click* on the **Ellipse** tool in the **Draw toolbox**. *Left-click* on the circle next to **Axis and Planar Rotation** in the **Ellipse Settings** dialogue box (Fig. 3.5), followed by *left-click* on **OK**.
7. *Left-click* on the **Ellipse** icon in the **Draw** toolbox.

   Ellipse Center of ellipse: *left-click* on 215,90
   Ellipse Axis endpoint: *left-click* on 285,90
   Ellipse Rotation around major axis: *left-click* on 240,120

   The ellipse appears on the screen. Note that this type of ellipse is based upon a circle as it is viewed from the front with its major axis horizontal and its minor axis vertical. As the circle is rotated around the major axis, it appears as an ellipse with a varying size minor axis. After the first two points have been picked, – for ellipse centre and axis endpoint, a line attached to the ellipse centre appears. The position of the end of this line determines the rotation of the ellipse around its major axis. The line can be *dragged* around the ellipse centre by moving the mouse.

8. *Double-click* on the **Ellipse** tool in the **Draw toolbox**. *Left-click* on the circle next to **Two Foci and Point** in the **Ellipse Settings** dialogue box, followed by *left-click* on **OK**.
9. *Left-click* on the **Ellipse** icon in the **Draw** toolbox.

   Ellipse First focus of ellipse: *left-click* on 360,180
   Ellipse Second focus: *left-click* on 360,80
   Ellipse Point on ellipse: *left-click* on 390,140

   The ellipse appears on screen. Note that this form of ellipse is based upon the construction in which the ellipse is a locus of a point based on the two foci. As the Point on ellipse is moved under mouse control, two locus lines attached to the foci appear, meeting on the ellipse at the cursor position. When the point on the ellipse has been selected by a *left-click* the ellipse is drawn on screen without the locus lines.

The reader is advised to practise drawing Arcs, Curves and Ellipses using the various tools described above.

Example 2 – Mirror from the Modify toolbox

1. **New ...** the file **A3SHEET.SKT**.
2. *Left-click* on the **Curve** icon in the **Draw** toolbox.

   Curve First point: *left-click* on 110,240
   Curve To point: *left-click* on 105,230
   Curve To point: *left-click* on 100,220
   Curve To point: *left-click* on 80,175
   Curve To point: *left-click* on 80,135
   Curve To Point: *left-click* on 95,100
   Curve To point: *double-click* on 105,90

3. *Left-click* on the **Polyline** icon in the **Draw** toolbox.

   Polyline From point: *left-click* on 100,240
   Polyline To point: *left-click* on 100,250
   Polyline To point: *double-click* on 130,250

   Left-click on the **Polyline** icon in the **Draw** toolbox.

   Polyline From point: *left-click* on 105,90
   Polyline To point: *left-click* on 105,90
   Polyline To point: *left-click* on 130,90

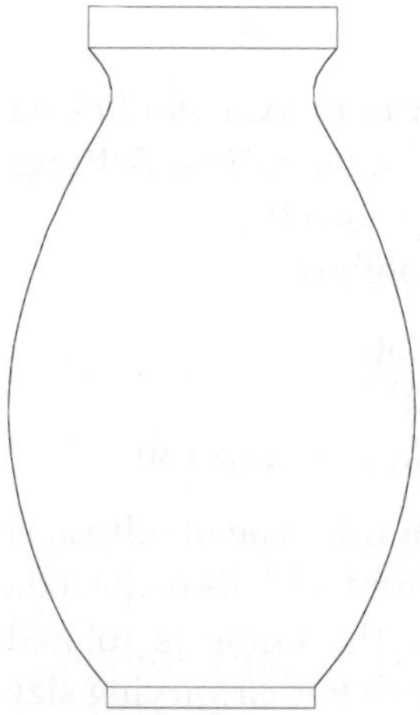

Fig. 3.6 Example 2

4. *Left-click* on the **Line** icon in the **Draw** toolbox.

   Line From point: *left-click* on 100,240
   Line To point: *left-click* on 130,240

   Left-click on the **Line** icon in the **Draw** toolbox.

   Line From point: *left-click* on 100,240
   Line To point: *left-click* on 130,240

5. *Left-click* on the **Mirror** icon (Fig. 3.7) in the **Modify** toolbox.

   Mirror Select object: *left-click* on 60,260
   Mirror Crosses/window corner: *left-click* on 145,75
   Mirror Base point: *left-click* on 130,250
   Mirror (Untitled): *left-click* on 130,85

Fig. 3.7 The **Mirror** tool icon

Example 3 – Mirror: a second example

The **Mirror** tool can save considerable time when constructing drawings which are symmetrical around both vertical and horizontal axes (Fig. 3.8). Work to the dimensions given using the tools **Polyline, Arc** and **Circle** and changing the **Line type** to **Centre** for the centre lines.

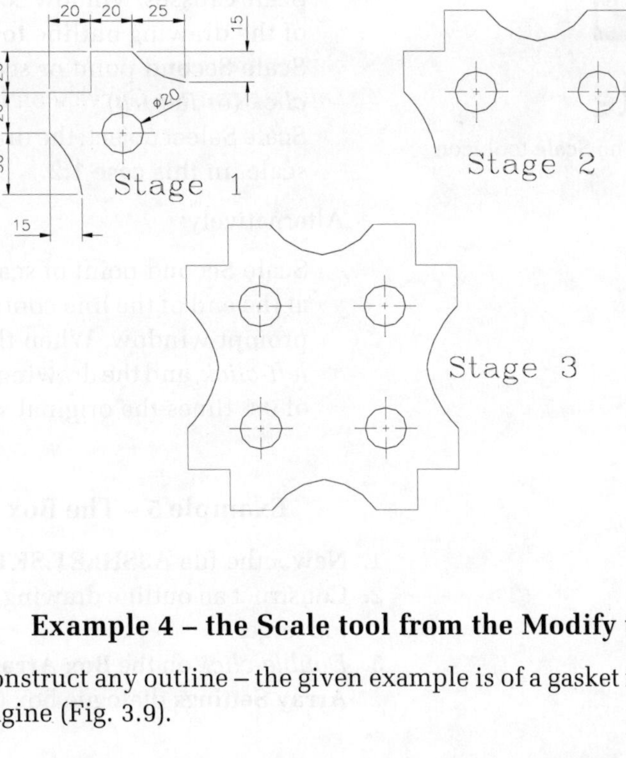

Fig. 3.8 Example 3

Example 4 – the Scale tool from the Modify toolbox

1. Construct any outline – the given example is of a gasket from a car engine (Fig. 3.9).

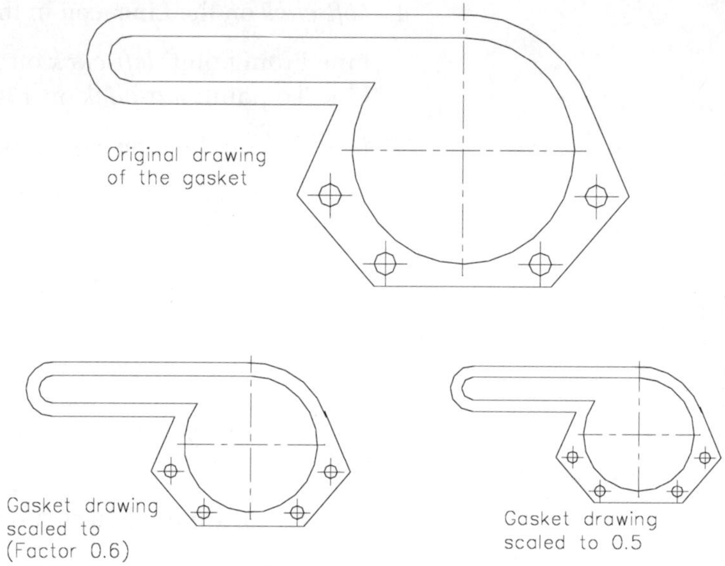

Fig. 3.9 Example 4

2. *Left-click* on the **Scale** icon (Fig. 3.10) in the **Modify** toolbox.

Fig. 3.10 The **Scale** tool icon

   Scale Select object: *left-click* above and to the left of the drawing outline
   Scale Crosses/window corner: *left-click* below and to the right of the drawing outline to highlight outline
   Scale Second point or scale factor: (Factor 0): *enter* 0.5 *right-click* (or *Return*)
   Scale Select object: the drawing outline re-draws to the desired scale, in this case 1:2

   Alternatively:

   Scale Second point of scale factor (Factor 0): move the cursor at the end of the line controlling the scale factor, watching the prompt window. When the (Factor 0) changes to (Factor 0.6) *left-click,* and the drawing outline re-draws to the desired scale of 0.6 times the original scale

Example 5 – The Box Array tool from the Edit toolbox

1. **New...** the file **A3SHEET.SKT**.
2. Construct an outline drawing – Fig. 3.11. Do not make your outline too large.
3. *Double-click* on the **Box Array** icon in the **Edit** toolbox. In the **Box Array Settings** dialogue box (Fig. 3.12) which appears:

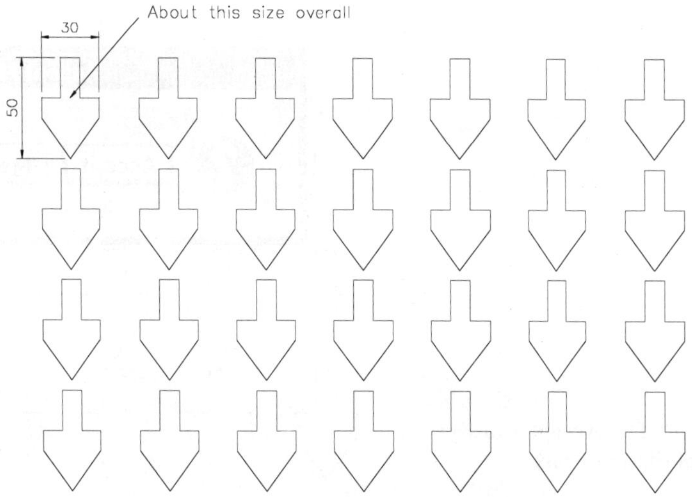

Fig. 3.11 Example 5

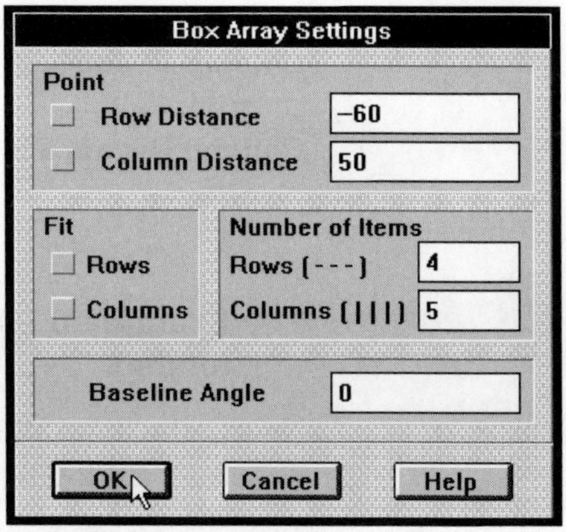

Fig. 3.12 The **Box Array Settings** dialogue box

In the box **Row Distance** *enter* -60
In the box **Column Distance** *enter* 50

In **Number of Items** box area:

In the box **Rows (– – –)** *enter* 4
In the box **Columns (| | |)** *enter* 5

Left-click on **OK**.

4. Box Array Select object: *left-click* above and to the left of your outline. Box Array Crosses/window corner: *left-click* below and to the right of your outline. The array appears on the screen, together with the **Accept/Modify/Cancel** dialogue box (Fig. 3.13). *Left-click* on **Accept** in this box, if satisfied with the array.

The sign conventions are as follows:

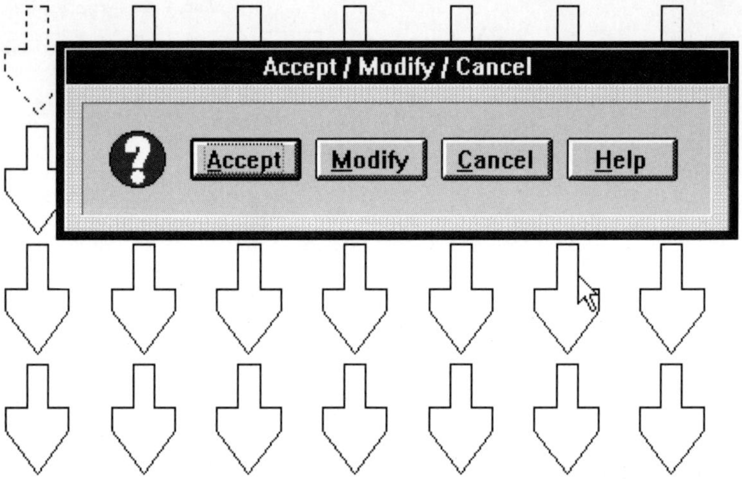

Fig. 3.13 The **Accept/Modify/Cancel** dialogue box

Row Distance is +ve if the rows are to be **above** the original.
Row Distance is –ve if the rows are to be **below** the outline.
Column Distance is +ve if the columns are to be to the **right** of the original.
Column Distance is –ve if the columns are to be to the **left** of the original.

Example 6 – using the Ring Array tool from the Edit toolbox

1. **New ...** the file **A3SHEET.SKT**.
2. Construct a drawing outline such as that shown in Fig. 3.14.

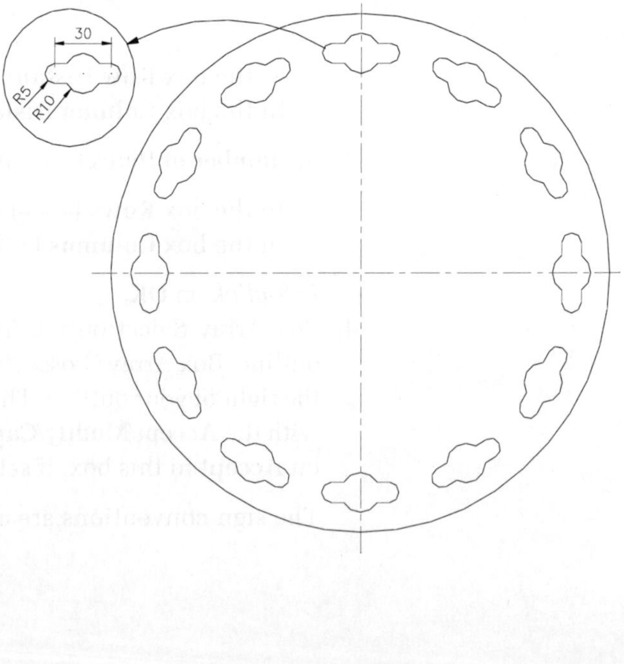

Fig. 3.14 Example 6

Fig. 3.15 The **Ring Array** tool icon

3. *Double-click* on the **Ring Array** icon (Fig. 3.15) in the **Edit** toolbox. In the appropriate boxes of the **Ring Array Settings** dialogue box (Fig. 3.16):

   Number of Items *enter* 12
   Included Angle *enter* 360 (if not already in the box)

   Degrees between items adjusts itself to the two numbers in the boxes above. Check that the tick in **Rotate Items as Copied** is present. Check that the tick in the **Point** box under **Center Point of Array** is present. *Left-click* on **OK**.

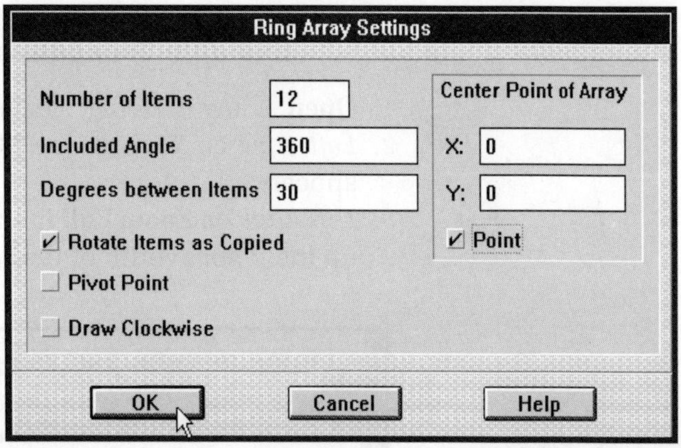

Fig. 3.16 The **Ring Array Settings** dialogue box

4. Ring Array Select object: *left-click* above and to the left of your outline. Ring Array Crosses/window corner: *left-click* below and to the right of your outline. Ring Array Center of Array: *left-click* on its centre point. *Left-click* on **Accept** in the **Accept/Modify/Cancel** dialogue box which appears on screen (Fig. 3.17).

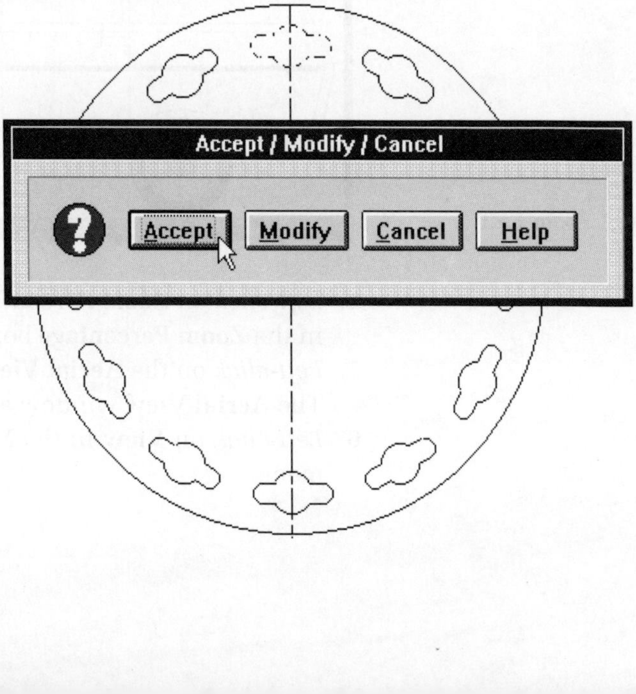

Fig. 3.17 The **Accept/Modify/Cancel** dialogue box

Note: The methods for forming a **Ring Array** are similar to those for forming a **Box Array**.

Zoom

The CAD zoom facility, allows the operator to examine the most minute areas of the drawing on screen. This enables accurate construction to be carried out in the smallest areas of the drawing and any modifications which may be needed in these small areas. AutoSketch for Windows includes a fast acting, efficient zoom facility.

Example 7 – Zoom and Zoom Box

1. **Open ...** any drawing. That shown here is a side view of a car.
2. *Left-click* on **View** in the **Menu bar** – the **View** pull-down menu appears.
3. *Left-click* on **Zoom Full** in the pull-down menu. The drawing takes up the whole width of the screen – Fig. 3.18.

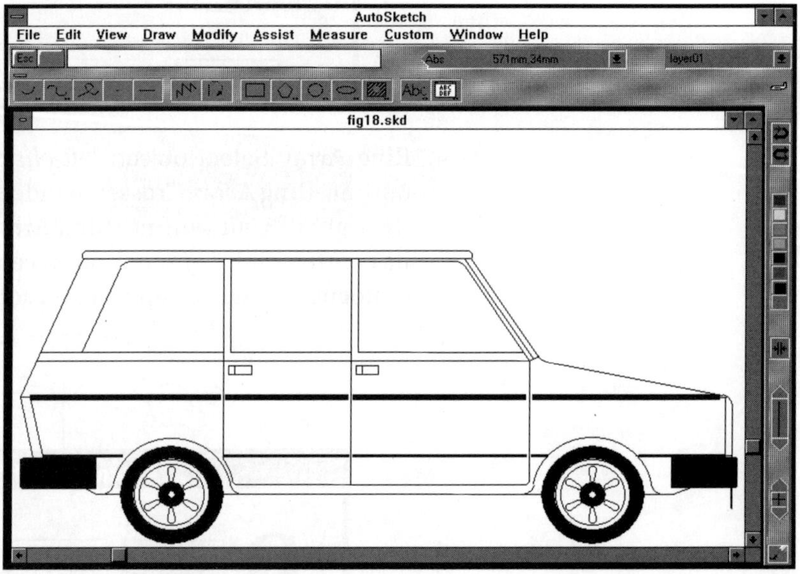

Fig. 3.18 A **Zoom Full**

4. *Left-click* on **Zoom Percent** in the **View** pull-down menu. *Enter* 50 in the **Zoom Percentage** box which appears (Fig. 3.19).
5. *Left-click* on the **Aerial View** button at the top right of the screen. The **Aerial View** window appears – Fig. 3.20.
6. *Left-click* on **View** in the **Menu bar**. *Left-click* on **Toolbox** in the menu.

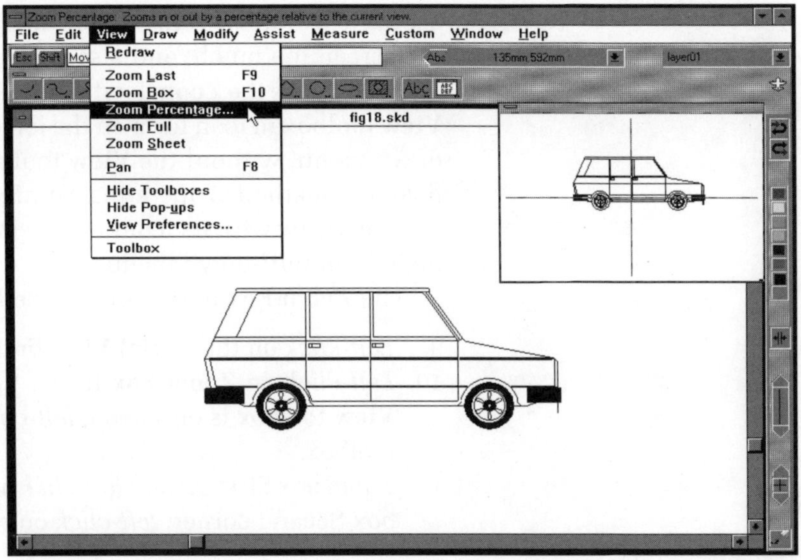

Fig. 3.19 A **Zoom Percent** (50) and the **View** pull-down menu

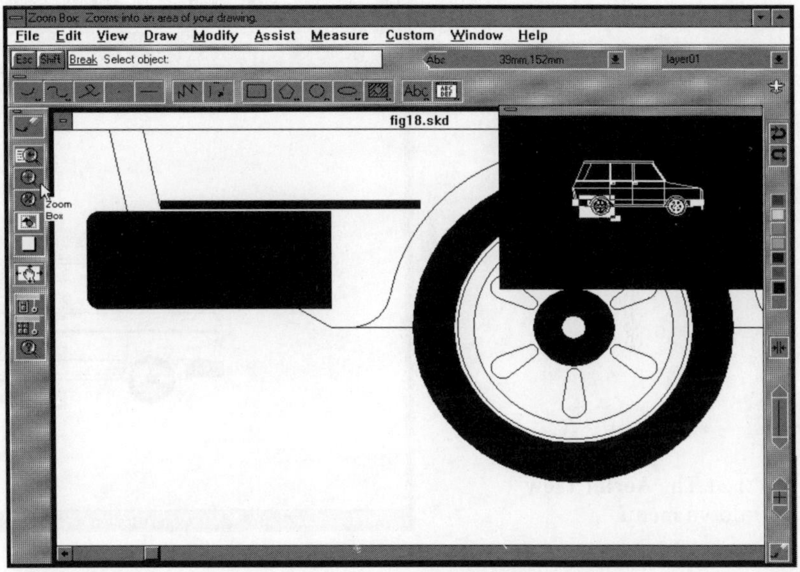

Fig. 3.20 A **Zoom Box** with the **Aerial View** window showing the zoom box area

7. *Left-click* on the **View** toolbox button. *Left-click* on **Copy Left** in the pull-down menu which appears. This places the **View** toolbox on the left of the screen.
8. *Left-click* on **Draw** in the **Menu bar** to restore the **Draw** toolbox to its original position at the top of the screen.

The operator has the choice between using the **View** pull-down menu for **Zooms** or using the **View** toolbox for **Zooms**. Personally I prefer using the pull-down menu in this case because it allows me access to a toolbox in the screen area – **Draw** – without cluttering up the screen too much, and also easy access to the **View** menu when wishing to use the **Zoom** facility. Figures 3.19 to 3.21 show both – the **View** toolbox in icon form on the left of the screen and the **View** pull-down menu, without the **View** toolbox permanently on screen. My favourite method of operating AutoSketch for Windows is to have the toolbox **Draw** with its icons on screen and to have access to all other tools from pull-down menus.

The **Zoom Box** operation proceeds as follows:

9. *Left-click* on the **Aerial View** button.
10. *Left-click* on **Zoom Box** in the **View** pull-down menu, or, if the **View** toolbox is on screen, *left-click* on the **Zoom Box** icon in the toolbox.
11. Zoom box First corner: *left-click* on required top left corner. Zoom box Second corner: *left-click* on required bottom right corner.

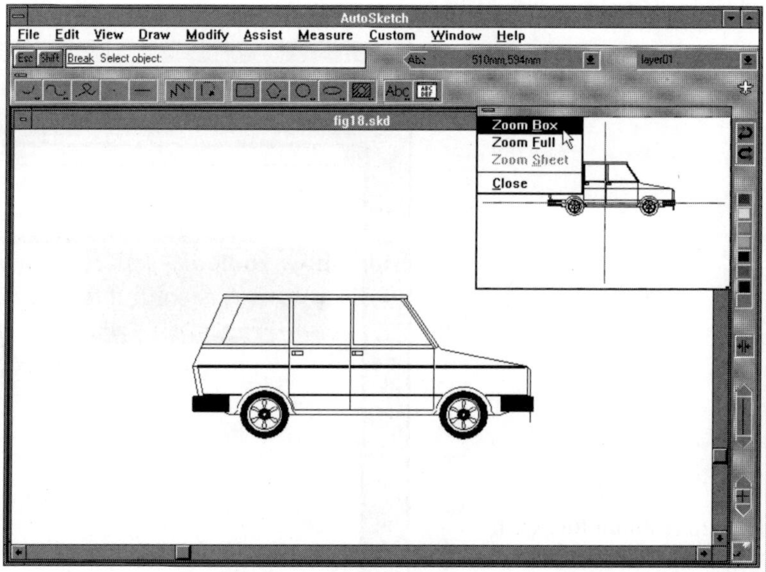

Fig. 3.21 The **Aerial View** pull-down menu

Note that the AutoSketch drawing screen changes to accommodate only the area of the drawing within the zoom box. At the same time, the Aerial View window changes to a reverse video display except for that part of the drawing which is within the zoom box – Fig. 3.20.

12. *Left-click* on **Zoom Full** in the **View** pull-down menu.
13. Try another **Zoom Box** in another part of the drawing. Watch the area in the Aerial View window which in not in reverse video. It now represents the area of the drawing showing the zoom box. The **Aerial View** window therefore lets the operator know which part of the drawing is being zoomed.

14. *Left-click* on each of the tools in the pull-down menu within the **Aerial View** window. If **Zoom Box** is chosen, the **Aerial View** window shows the area covered by the **Zoom Box** on screen. If **Zoom Full** is chosen, the Drawing changes to occupy as much space as it is able to in the **Aerial View** window. *Left-click* on **Zoom Sheet** and the drawing reverts to its size controlled by the **Sheet** settings.

Notes on tools

The Aerial View pull-down

Zoom Box – view will be of the **Zoom Box** only (Fig. 3.20).
Zoom Full – view as large as the window allows.
Zoom Sheet – view repeats the **Zoom Sheet** of the screen to a smaller scale to fit in the window.

View toolbox

Zoom X – File Factor appears in the Prompt bar and a dialogue box **Zoom Percent** appears on screen. *Enter* the required factor figure, either smaller or larger than 1, and the drawing re-draws at the stated Factor number. Note that the **Zoom Percent** number is with reference to the current zoom on screen, not to the scale of the original drawing.
Hide/Show Toolbox – *left-click* on this tool and all toolboxes on screen disappear. A second *left-click* and the toolboxes reappear.
Pan – *left-click* on **Pan** and the screen with its drawing can be panned: *left-click* on any point on the screen, followed by *left-click* on another point on screen. The drawing and screen move, in other words the coordinate system moves by the distance between the two *left-clicks*.
Zoom Pref – brings up the **View Preferences** dialogue box (not illustrated here). The reader is advised to experiment with the settings in this dialogue box.

Other tools

Ortho-Assist toolbox – when this tool is selected, lines, polylines etc. can only be drawn horizontally or vertically on screen.
Attach-Assist toolbox – **Attach** points allow the cursor to snap on the points of objects defined by the **Attach** tool selected. *Note:* **Attach** *must* itself be on before any other **Attach** tool will function.
Node – to a **Point** drawn on screen or to the node of text (the **Justification** point of the text as set in the **Text Settings** dialogue box) already on screen nearest to the point selected.

End point – to the end of the object on screen nearest the selected point.
Mid point – to the middle of the object on screen nearest to the selected point.
Perpendicular – the object being drawn attaches itself as a perpendicular to the nearest object on screen nearest to the selected point.
Intersect – to the point of intersection between two objects on screen nearest to the selected point.
Centre – to the centre of the arc or circle on screen nearest to the selected point.
Quadrant – to the quadrant point of an arc or circle on screen nearest to the selected point.
Tangent – to the tangential point of the circle, arc or curve on screen nearest to the selected point.

The Escape button

If a tool has been selected in error, or if a start has been made with a tool in a wrongly selected position, there are three ways of rectifying the error:

1. Press the **Esc** key of the computer's keyboard.
2. *Left-click* on the **Esc** button in the AutoSketch drawing screen.
3. *Left-click* on the tool icon, or on another tool icon if the wrong tool has been selected.

Exercises

1. Fig. 3.22. Construct the drawing of an engine gasket shown without including the dimensions. Use **Copy** or **Multiple Copy** for some parts of your drawing.

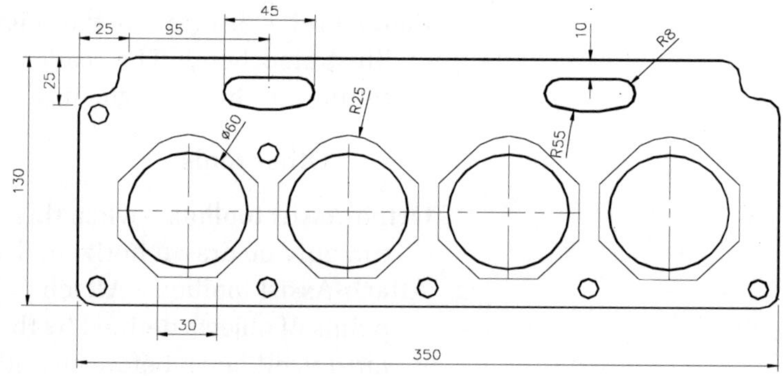

Fig. 3.22 Exercise 1

2. Fig. 3.23. Construct the given drawing. Note the use of **Polar** coordinates for drawing the lines limiting the outline of the part to be acted in by **Box Array**. Do not include the dimensions.

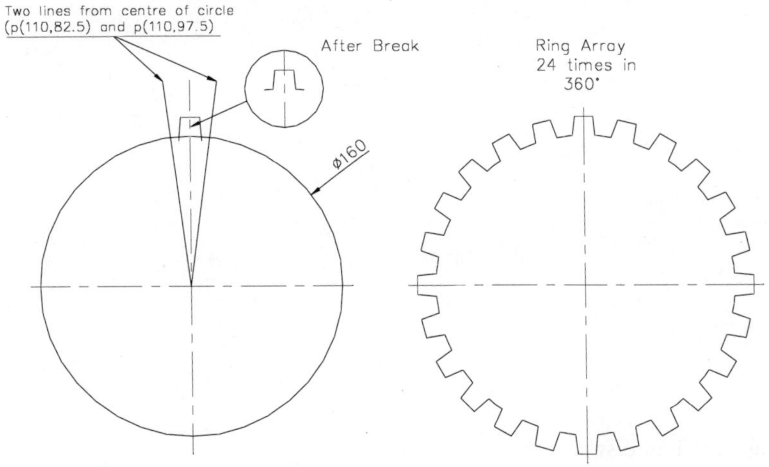

Fig. 3.23 Exercise 2

3. Fig. 3.24. Construct the two **Polar Arrays**.

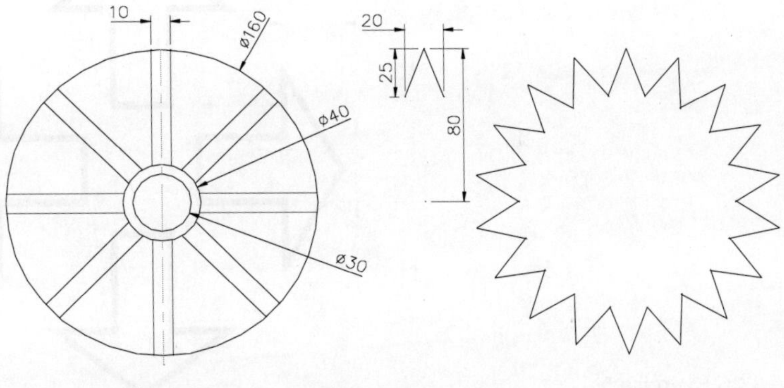

Fig. 3.24 Exercise 3

4. Fig. 3.25. Using **Line**, **Circle** and **Box Array** copy the given drawing.
5. Fig. 3. 26. With the aid of the tools named in the drawing copy the given drawing. Do not include any dimensions.
6. Fig. 3.27. Some ideas for drawings mainly constructed with the **Ellipse** tool from the **Draw** toolbox. Use any suitable sizes for your drawings.

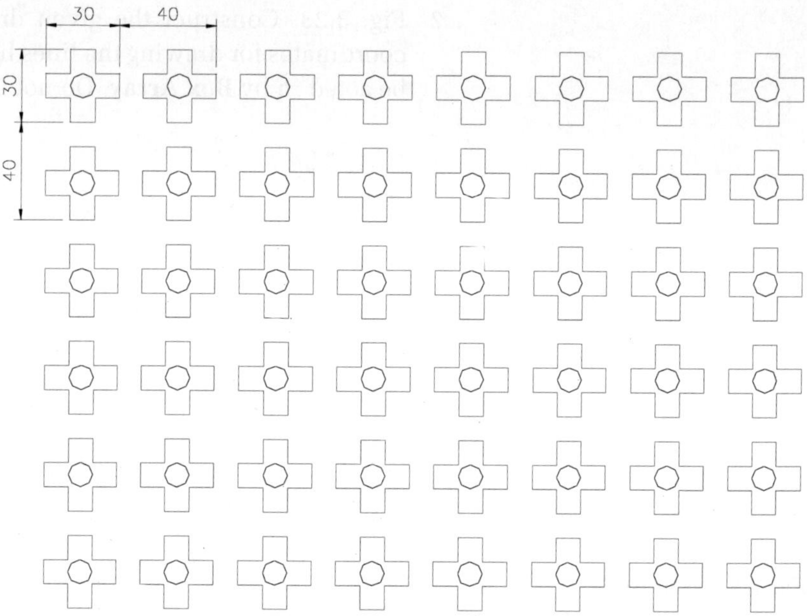

Fig. 3.25 Exercise 4

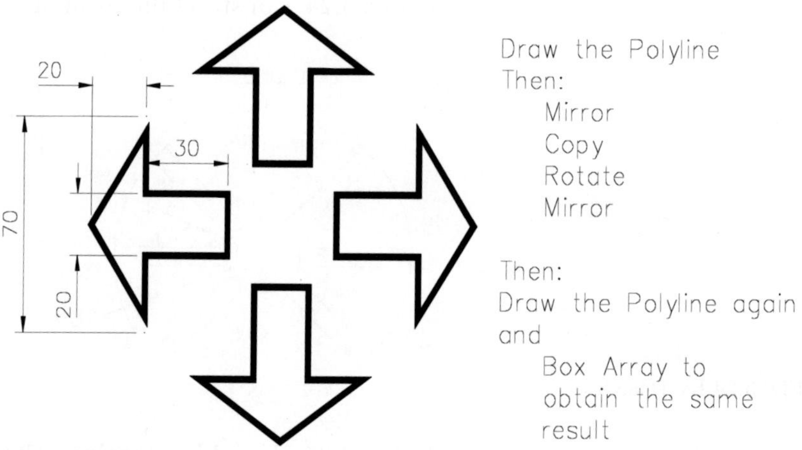

Fig. 3.26 Exercise 5

Fig. 3.27 Exercise 6

CHAPTER 4

Orthographic projection, layers, dimensions, text

Theory of orthographic projection

In order to convey accurate visualisation of the shape of three-dimensional articles as drawings on a sheet of paper, which is itself two-dimensional, the geometrical system known as **orthographic projection** is commonly employed. This form of projection depends upon placing the article to be drawn in an area bounded by flat planes. Usually one or more **vertical planes (VP)** and a **horizontal plane (HP)** are involved, although orthographic projections more advanced than those given in this book may well use planes placed at other angles in space. As far as the draughtsperson is concerned the planes are invisible, but it is as well, when constructing orthographic projections, to appreciate the underlying theory of planes to assist in achieving accurate drawings.

If a vertical and a horizontal plane cross one another as in Fig. 4.1, they must be at right angles and are said to be orthogonal to each other. When viewed from a side, it will be seen that the crossing

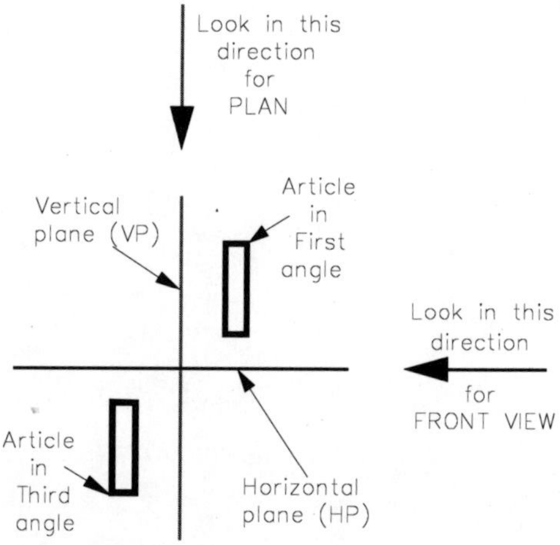

Fig. 4.1 VP and HP as seen from a side forming four dihedral angles

planes embrace four right angles. Reading anti-clockwise from top right, these right angles are known as the first angle, second angle, third angle and fourth angle. It is only the first and third angles which are used in orthographic projection.

The article to be drawn is imagined as being placed in either the first or in the third angle. It is then looked at from the front and/or from above. In first angle projection what is seen is cast onto the vertical plane behind the article and onto the horizontal plane below the article (Fig. 4.2).

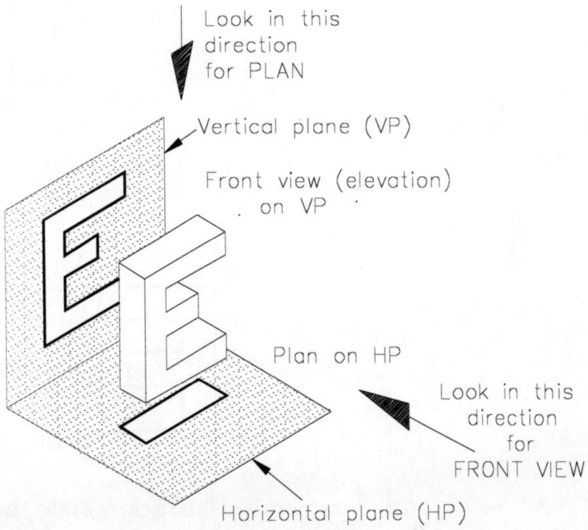

Fig. 4.2 The theory of first angle projection

In third angle projection what is seen is as if viewed through the planes and is drawn onto the vertical plane in front of the article and onto the horizontal plane above the article (Fig. 4.3).

The horizontal plane, with its plan outline, is then rotated through 90° so as to be in line with the vertical plane, forming a single two-dimensional plane – the drawing sheet. Finally, since the HP and VP are imaginary, they are not included with the drawing. Symbols to represent first or third angle drawing are often included with orthographic projections to indicate to the reader of a drawing in which angle they have been drawn (Fig. 4.4).

Example 1 – a simple first angle orthographic drawing

1. **New...** the file **A3SHEET.SKT**.
2. Make sure that **Grid** (set to 10 units) and **Snap** (set to 5 units) are on.

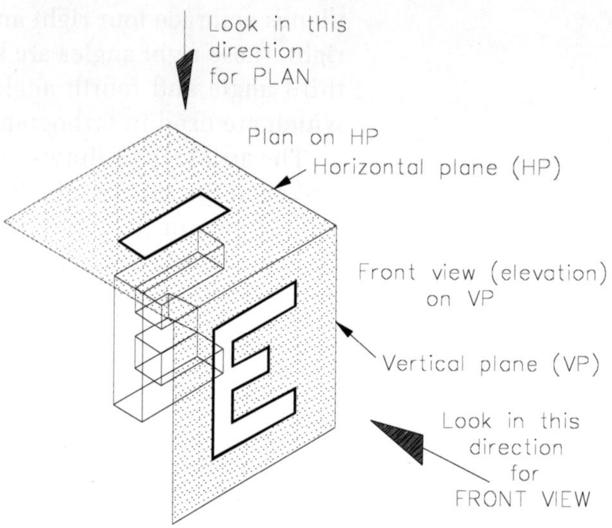

Fig. 4.3 The theory of third angle projection

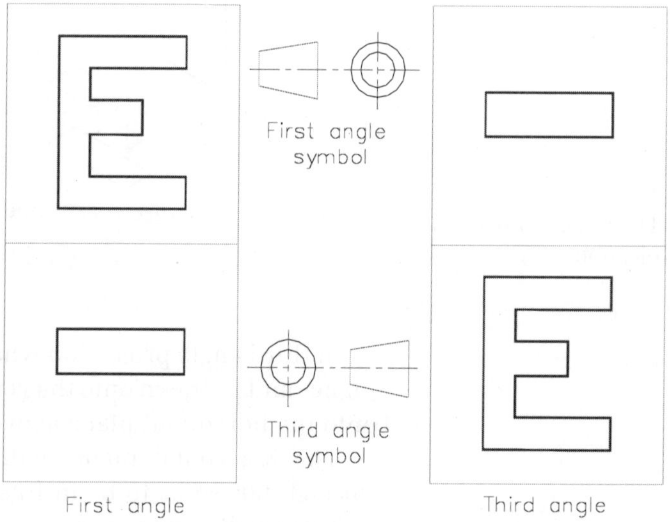

Fig. 4.4 Front view and plane in first angle and in third angle projection, together with the symbols of projection

3. Without including any of the dimensions and working with the tools **Line**, **Circle**, **Arc** and **Fillet**, construct the given front view and plan (Fig. 4.5) of a half bearing block. You will find that the tool **Ortho** and use of *left-click* on the cursor icon (the cross in the **Vertical Status bar**) will assist in obtaining accurate projection lines between the two views.
4. Add centre lines and hidden detail lines to your views and the symbol for first angle projection.

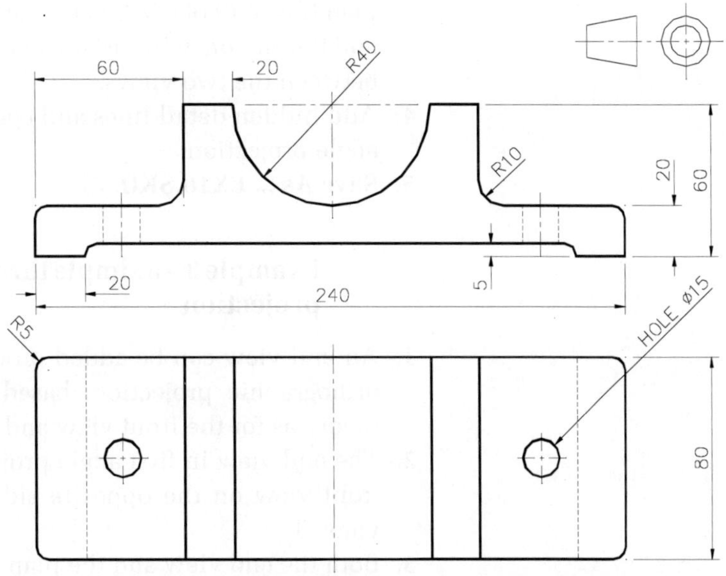

Fig. 4.5 Example 1

Example 2 – a simple third angle orthographic projection

1. **New...** the file **A3SHEET.SKT**.
2. Make sure that **Grid** (set to 10 units) and **Snap** (set to 5 units) are on.
3. Without including any of the dimensions and working with the tools **Line**, **Circle**, and **Fillet**, construct the given front view and

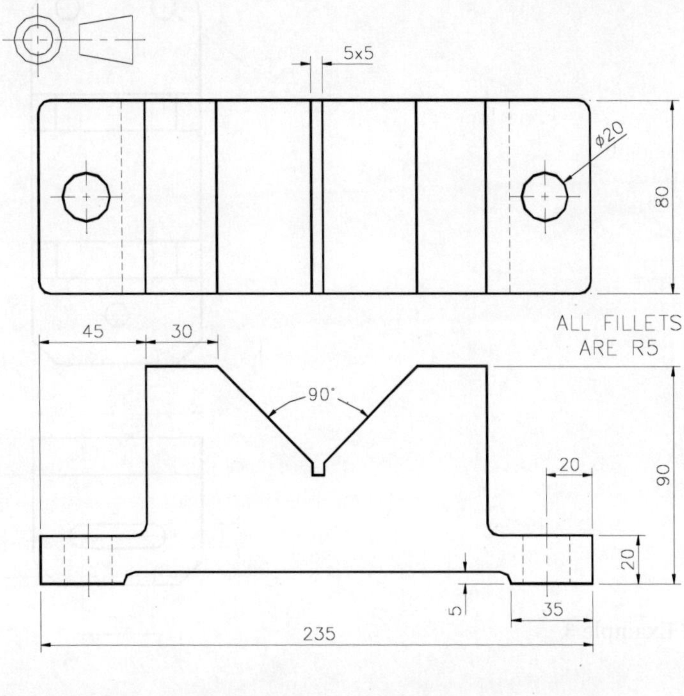

Fig. 4.6 Example 2

plan (Fig. 4.6) of a V-block. The **Ortho** tool from the **Assist** toolbox and the cursors will assist you in obtaining correct projection lines between the two views.

4. Add hidden detail lines and centre lines and the symbol for third angle projection.
5. **Save As ... EX16.SKD**.

Example 3 – a simple three-view first angle orthographic projection

1. An end view can be added on either side of the front view of an orthographic projection, based on exactly the same projection theory as for the front view and plan (Fig. 4.7).
2. The end view in first angle projection is placed on the side of the front view on the opposite side from which the object is being viewed.
3. Both the end view and the plan face outwards from the front view – their front surfaces are towards the edges of the drawing sheet.
4. Make sure that **Grid** (set to 10 units) and **Snap** (set to 5 units) are on. Without including any of the dimensions and working with the tools **Line**, **Circle**, and **Arc**, construct the given front view, end view and plan of a RACK. You will find that the tool **Ortho** and *left-click* on the cursor icon (the cross in the **Vertical Status bar**) – will assist

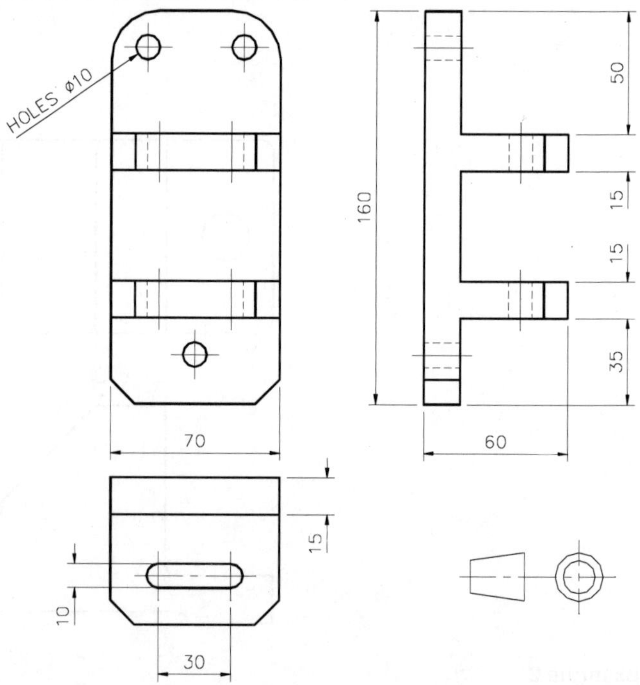

Fig. 4.7 Example 3

in obtaining accurate projection lines between the two views. Add centre lines and hidden detail lines to your views and the symbol for first angle projection.

5. **Save As ... EX17.SKD**

Example 4 – a three-view third Angle projection

1. An end view can be added on either side of the front view of an orthographic projection, based on exactly the same projection theory as for the front view and plan (Fig. 4.8).
2. The end view in third angle projection is placed on the same side of the front view as that from which the object is being viewed.
3. Both the end view and the plan face inwards towards the front view – their front surfaces are facing inwards away from the edges of the drawing sheet.
4. Make sure that **Grid** (set to 10 units) and **Snap** (set to 5 units) are.
5. Without including any of the dimensions and working with the tools **Line**, **Circle**, and **Arc** and making use of the **Tangent** attach tool from the **Assist** toolbox, construct the given front view, end view and plan of a holder and pin. You will find that the tool **Ortho** and using *left-click* on the cursor icon (the cross in the **Vertical Status bar**) will assist in obtaining accurate projection lines between the two views. Add centre lines and hidden detail lines to your views and the symbol for third angle projection.

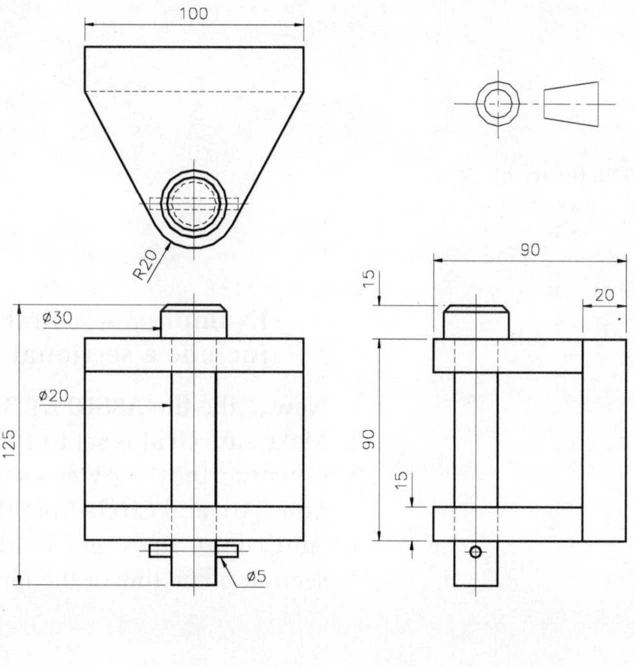

Fig. 4.8 Example 4

Sectional views in orthographic projections

A sectional view is a view of the cut surface of an item in an orthographic projection. Not only is the cut surface shown in the section, but all details behind the cut surface should be included in the view.

It is not common practice to include hidden detail in sectional views.

Sectional views in engineering drawings are usually 'hatched' with lines at 45° spaced about 3 to 4 mm apart (Fig. 4.9). The edge of the cutting plane is usually indicated with a section plane line – a centre line ending in two short thick lines. If it is obvious where the section plane lies there is no real need to include the cutting plane line.

Sectional views are usually labelled to indicate to which section plane they belong.

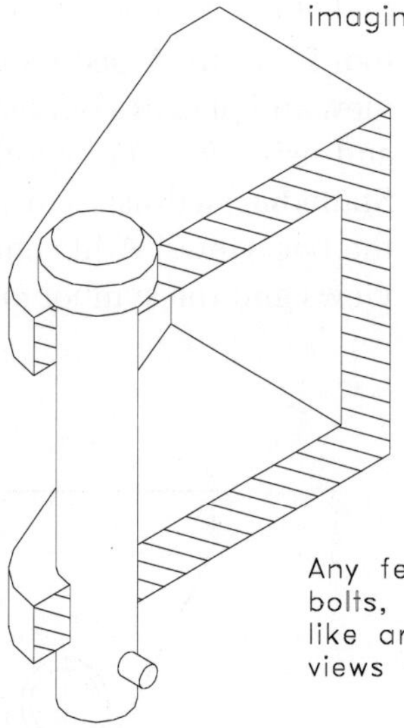

Fig. 4.9 The theory of sectional views

Example 5 – a first angle orthographic projection to include a sectional view

1. **New ...** the file **A3SHEET.SKT**.
2. Make sure **Grid** is set to 10 and **Snap** set to 5.
3. Construct the three views shown in Fig. 4.10 with the aid of the tools **Line**, **Arc** and **Circle** from the **Draw** toolbox.
4. Add centre lines and hidden detail lines. Do **NOT** include the section plane line or the letters A and A–A.

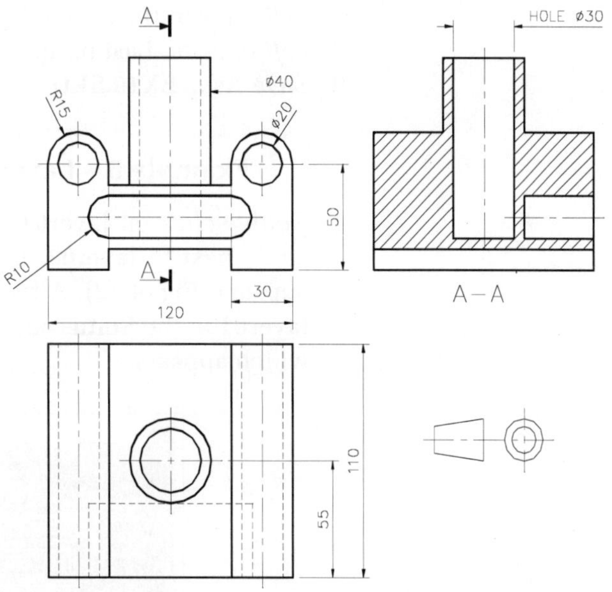

Fig. 4.10 Example 5

5. *Double-click* on the **Pattern Fill** icon in the **Draw** toolbox. (Fig. 4.11 shows the **Pattern Fill Accept/Modify/Cancel** dialogue box.) *Double-click* on **[..]** in the directory box of the **Pattern Settings** dialogue box. *Left-click* on the **ansi31** in the changed dialogue box. *Enter* 0 against **Angle**. *Enter* 2 against **Scale**. *Left-click* on **OK**.
6. *Left-click* on **View** in the **Menu bar**. *Left-click* on **Zoom Box** in the **View** pull-down menu. **Zoom** the un-hatched sectional end view in a window.

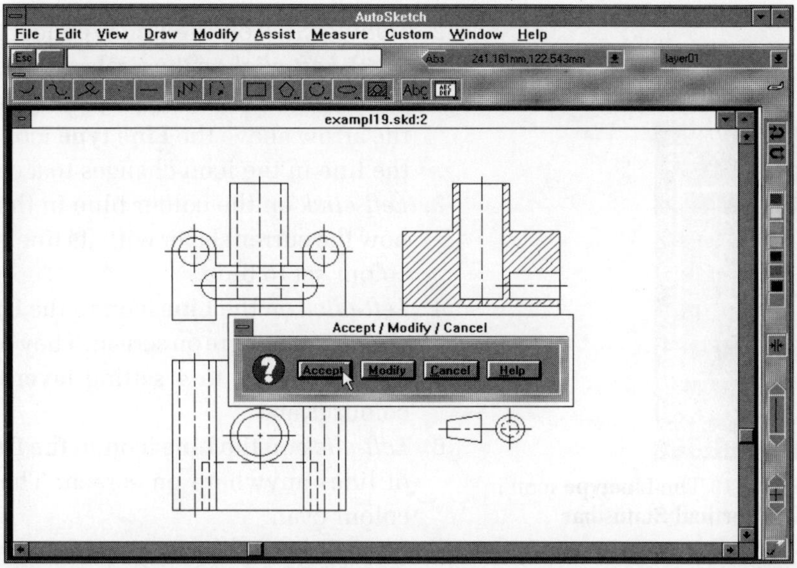

Fig. 4.11 The **Accept/Modify/Cancel** dialogue box in connection with **Pattern Fill**

7. *Left-click* in the area to be hatched.
8. *Left-click* on **Last** in the **View** pull-down menu.
9. **Save As ... EX19.SKD**.

Example 6 – Layers

1. *Double-click* on **layer01** icon in the **Status bar**. *Left-click* on the circle next to **layer02** in the **Layer Settings** dialogue box which appears (Fig. 4.12). Alternatively, *left-click* on the arrow next to **layer01** in the **Status bar**, then *left-click* on **layer02** in the list box which appears.

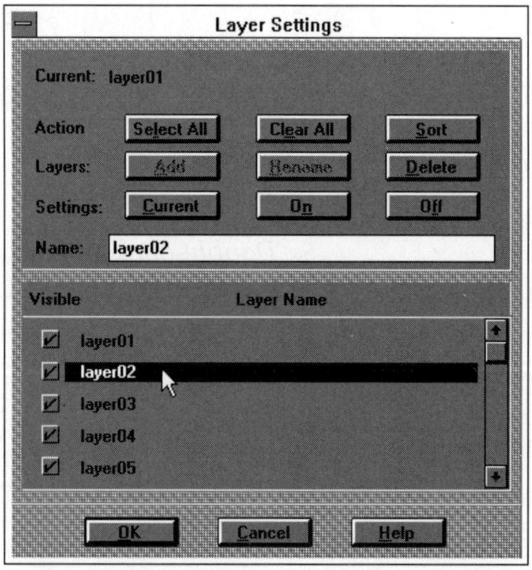

Fig. 4.12 The **Layer Settings** dialogue box

2. *Double-click* on the **Linetype** icon in the **Vertical Status bar** (Fig. 4.13). *Left-click* on the circle next to **Center** in the **Line Type Settings** dialogue box which appears (Fig. 4.14). Alternatively, *left-click* on the arrow above the **Line type** icon in the **Vertical Status bar** until the line in the icon changes to a centre type.
3. *Left-click* on the colour blue in the **Vertical Status bar**. **layer02** is now the current layer with its line type set as centre and its drawing colour set to blue.
4. *Left-click* on the **Line** icon in the Draw toolbox and draw a number of lines anywhere on screen. They will be centre lines of colour blue.
5. Repeat steps 1 to 4 setting **layer03** with **Line Type Hidden** and colour cyan.
6. *Left-click* on the **Line** icon in the **Draw** toolbox and draw a number of lines anywhere on screen. They will be hidden type lines of colour cyan.

Fig. 4.13 The **Linetype** icon in the **Vertical Status bar**

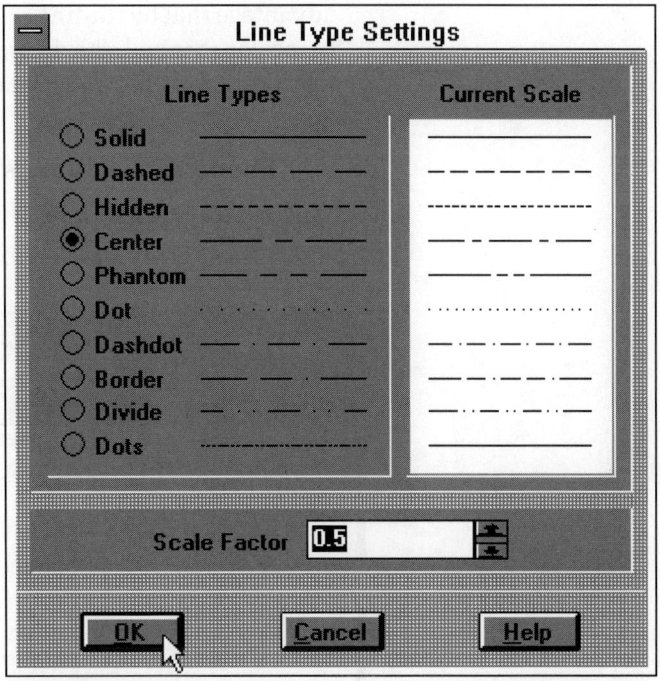

Fig. 4.14 The **Line Type Settings** dialogue box

Line Scale Factor

The **Scale Factor** figure determines the lengths and spaces between lines which consist of dots and/or dashes. The default scale 1 applies to a **Limits** setting of *x,y* = 420,297 at the top right of the AutoSketch screen. With other **Limits** settings the **Scale Factor** must be reset on the basis that **Scale Factor** 1 is the setting when **Limits** are set to 420,297. Personally I prefer the setting to be 0.5, even with a 420,297 template.

Notes on layers

1. Only 10 layers can be set in AutoSketch for Windows.
2. The **Layers Settings** list box shows a series of ticks in buttons against the Layer numbers. When a tick is showing in a button it means that the layer to which the button refers is on and its contents are visible on screen.
3. If there is not a tick in the button against a layer number the layer to which the tick refers is off and its contents become invisible on screen.
4. On start-up, the default layer is number 1.
5. When constructing a complicated drawing, it is advisable to place different features of the drawing on different layers. This has the

advantage that by turning off those layers on which features are not being constructed, the drawing becomes less cluttered, making the addition of further construction easier.

Example 7 – dimensioning

1. **New ... A3SHEET.SKT.**
2. Draw the outlines as shown in Fig. 4.15 to any suitable sizes.

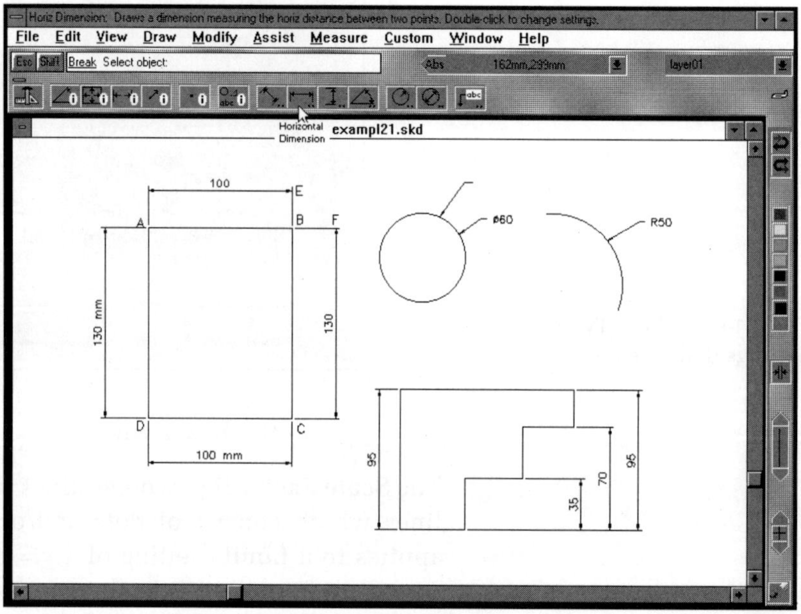

Fig. 4.15 Example 7

3. Make sure that **Grid** and **Snap** are on.
4. *Double-click* on the **Text Edit** icon in the **Draw** toolbox and check that the style of text is **simplex**.
5. *Left-click* on **Measure** in the **Status bar**.
6. *Left-click* on the **Leader** icon in the **Measure** toolbox. *Left-click* on each of the circles to left and right of **Solid** in the **Dimension Settings** dialogue box (Fig. 4.16). Check that a tick is in the box next to **Aligned**. *Enter* 5 in the **Options Text Size** box. *Enter* 4 in the **Options Arrow Size** box. *Left-click* on **OK**.
7. *Left-click* on the **Horizontal Dimension** icon in the **Measure** toolbox.

Horizontal Dimension From point: *left-click* on corner A
Horizontal Dimension To point: *left-click* on corner B
Horizontal Dimension, Dimension line location: *left-click* near E

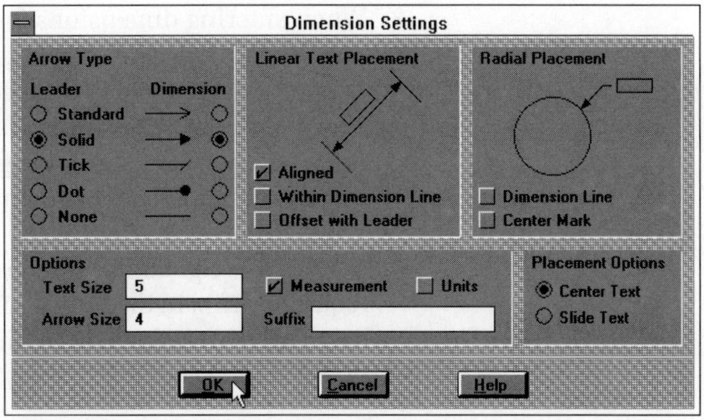

Fig. 4.16 The **Dimension Settings** dialogue box

8. *Left-click* on the **Vertical Dimension** icon in the **Measure** toolbox.

   Vertical Dimension From point: *left-click* on corner B
   Vertical Dimension To point: *left-click* on corner C
   Vertical Dimension, Dimension line location: *left-click* near F

9. In a similar manner dimension the outline in the bottom right of the screen.
10. Press function key **F10** – this calls the tool **Zoom Box**. Window the circle on screen in a zoom window.

    Left-click on the **Leader** icon in the **Measure** toolbox
    Left-click on the **Snap** icon in the **Assist** toolbox to turn **Snap** off
    Left-click on the **Leader** icon in the **Measure** toolbox
    Leader Start point: *left-click* on a point on the circle circumference
    Leader To point: *left-click* on required point
    Leader Text Location: *left-click* at a suitable point – the **Text Editor** dialogue box appears

    Enter space %%c60 followed by a *left-click* on **OK**.
11. In a similar manner dimension the arc, entering R60 in the **Text Editor** dialogue box.

Notes on dimensioning

1. The style of the dimension text is set in the **Text Settings** dialogue box.
2. If the bonus disk from Autodesk has been loaded into the directory holding the AutoSketch for Windows files, circles and arcs can be automatically dimensioned with the **radial** toolbox files from the bonus disk. To use the tools in this toolbox select **Open Toolbox** from the **Utility** pull-down menu, followed by selection of **radial.txt** from the file list that appears on screen.

3. When entering dimensions for arcs or circles, enter a space before the text, otherwise the text appears on screen right up against the line determining the text position.
4. Note the method of entering the diameter symbol with %%c. Other symbols are %%d for the degree symbol and %%p for the plus/minus tolerance symbol when including toleranced dimensions in drawings.
5. If a suffix – e.g. mm – is to be entered after a dimension, *enter* the required suffix in the **Suffix** box of the **Dimension Settings** dialogue box The required suffix is then automatically entered with each dimension.

Exercises

Dimensions have been included in many of the exercises in earlier chapters. Go back to these by a *left-click* on **Open ...** in the **File** pull-down menu and open any of the drawings you have made in answer to earlier exercises. Now add dimensions to your drawings.

Here are some additional exercises for practising adding dimensions to a drawing.

1. Fig. 4.17. **New ... A3SHEET.SKT**. Construct the outlines with the aid of the **Polyline** tool. Then add dimensions as indicated.

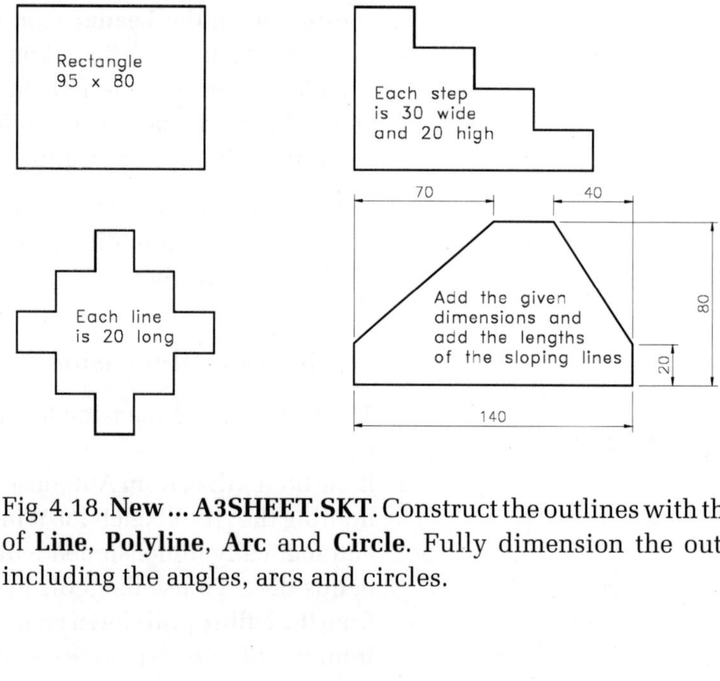

Fig 4.17 Exercise 1

2. Fig. 4.18. **New ... A3SHEET.SKT**. Construct the outlines with the aid of **Line**, **Polyline**, **Arc** and **Circle**. Fully dimension the outlines including the angles, arcs and circles.

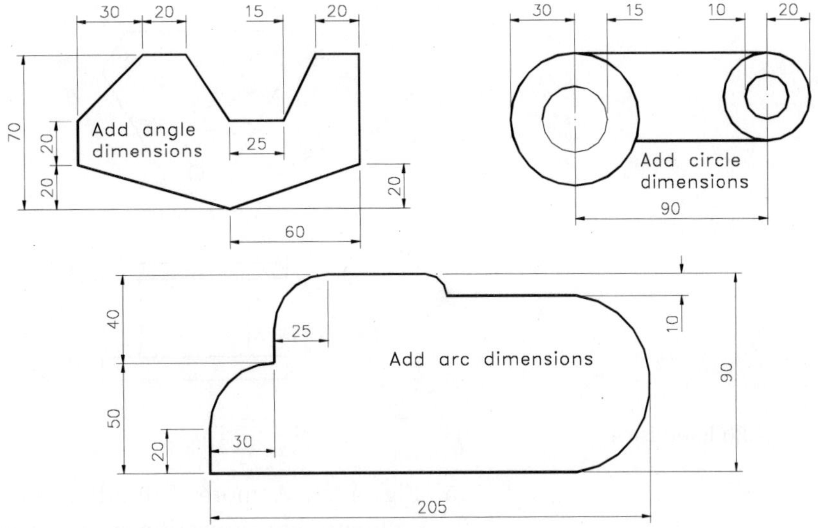

Fig. 4.18 Exercise 2

3. Fig. 4.19. **New ... A3SHEET.SKT**. Construct the outline with the aid of **Polyline**. *Enter* %%p0.05 in the **Suffix** box of the **Dimension Settings** dialogue box. Then add dimensions to the outline you have just drawn, but including the tolerance ± 0.05.

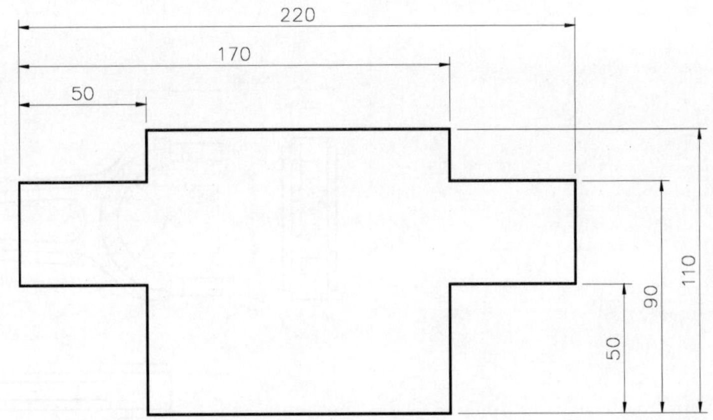

Fig. 4.19 Exercise 3

4. Fig. 4.20. A three-view third angle orthographic projection of a connector is given. Draw the front view and plan in first angle projection and add a sectional end view with the section plane running centrally through the hole in the component.

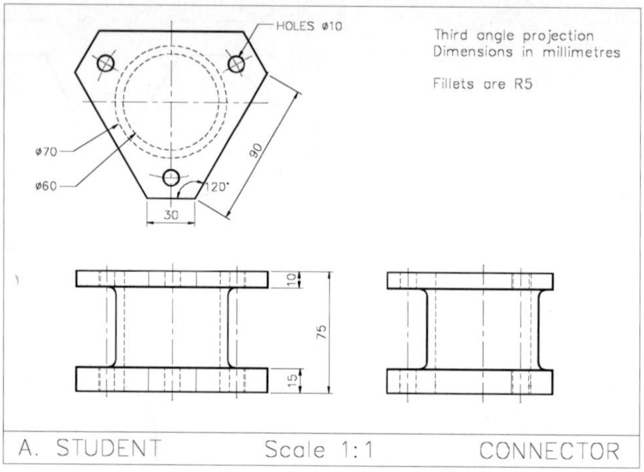

Fig. 4.20 Exercise 4

5. Fig. 4.21. A more difficult exercise. Copy the given four-view orthographic projection of a connecting rod. In order to achieve accurate outlines for the four views, you will need to make good use of the **Attach** tools – **End**, **Intersection** and **Quadrant** in particular. Text is either 6 units or 10 units high. Another tool which would help in ensuring correct projection between views is **Ortho**.

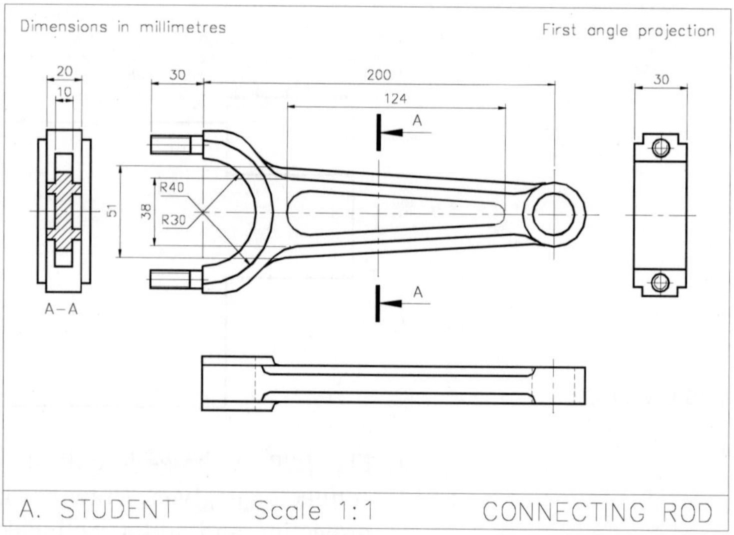

Fig. 4.21 Exercise 5

6. Fig. 4.22. Although this exercise is, again, more difficult than previous ones, by now if all the work described in the book so far has been followed, this exercise should not present too many difficulties. The front and end views of a small house are given. The drawing is intended to be constructed in your **A3SHEET** template. The given

Fig. 4.22 Exercise 6

dimensions refer, of course, only to the drawing and not to the actual dimensions of the house. Separate dimensioned drawings of a window, the house door and the chimney are also given in Fig. 4.23. Some hints to help the construction of the two views are:

(a) Set **Snap** to 1 unit and construct the windows, chimney and door in zoom boxes.

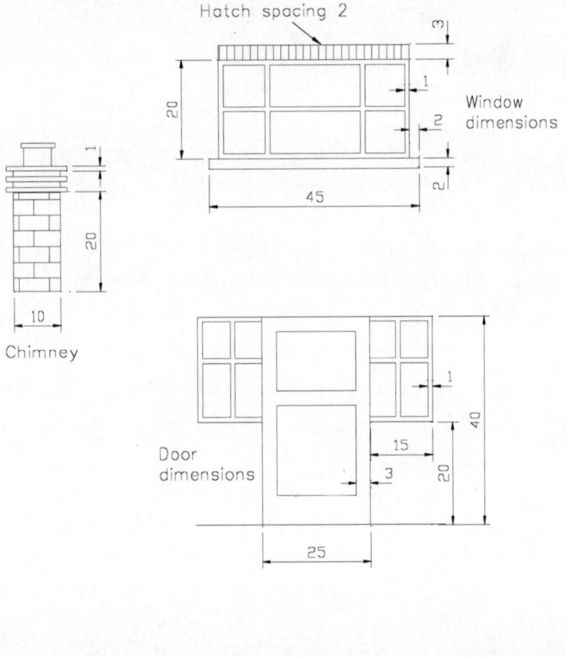

Fig. 4.23 Exercise 6

(b) The **Pattern Settings** are:

Walls – pattern **bricks**, **Angle 0**, **Scale 20**
Roof – pattern **ar-b88**, **Angle 0**, **Scale 0.7**

(c) Make full use of **End Attach** when adding the **Pattern Fills**.
(d) Use **Mirror** to copy one side of the door window to the other side.
(e) Use **Multiple Copy** to copy one window to obtain the others.
(f) Reset **Snap** to 5 units after constructing the windows.

CHAPTER 5

Revision and more about some tools

Introduction

In this chapter we will deal with some features which have not so far been mentioned and will include examples of more advanced constructions with tools already described in simpler examples in previous pages. Some more advanced exercises are also included at the end of the chapter.

Function key shortcuts

Construction time can be quite considerably reduced by using keyboard shortcuts. There are three groups of these shortcuts:

1. Pressing a function key – on marked **F1** to **F12**.
2. Pressing the **Alt** key and a function key.
3. Pressing the **Ctrl** key plus a function key.

Table 5.1 lists the keyboard shortcuts. When AutoSketch for

Table 5.1 AutoSketch for Windows keyboard shortcuts

Function key	*Functions*		
	F only	*Alt+F*	*Ctrl+F*
F1	Help	Line	Arc Mode
F2	Undo	Redo	Box Array
F3	Erase	Arc	Mirror
F4	Break	Exit	Ring Array
F5	Move	Ortho	Rotate
F6	Copy	Grid	Scale
F7	Stretch	Snap	Box
F8	Pan	Attach	Ellipse
F9	Last view	Group	Pattern Fill
F10	Zoom box	Ungroup	Macro input
F11	Print	—	—
F12	Save	—	—

Windows is first purchased, a card slip with the keyboard shortcuts printed on its surface is supplied with the disks and manuals. This slip is designed to be taped to the keyboard just above the function keys to indicate the key shortcuts to the operator. Personally, I find this slip allows me to use the keyboard shortcuts without having to memorise their meanings. Construction speed is greatly enhanced if the operator takes advantage of the supplied slip and uses the keys rather than the mouse to select tools and commands.

Text options

A large variety of different styles of text fonts are available in AutoSketch for Windows. Some of these have already been mentioned and the **Text Settings** dialogue box on page 11 shows most of the styles available.

Fig. 5.1 shows a number of examples of different styles of text of various heights and obliquing angles.

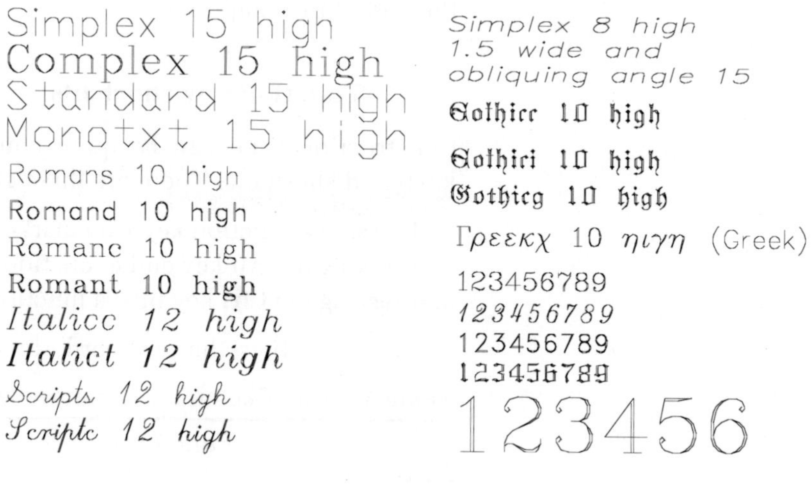

Fig. 5.1 A variety of fonts from AutoSketch

Example 1 – the Change Properties tool

1. *Left-click* on **Edit** in the **Status bar**. *Left-click* on **Change Properties** in the **Edit** pull-down menu (Fig. 5.2).
2. Property Select object: *left-click* on the object whose property is to be changed. The change takes place.

The change will be to the current settings for the object. Thus before the change from **Polyline** from 10 wide and solid filled to 5 wide and blank filled can take place, the settings for **Polyline** in the **Polyline Settings** dialogue box must first be set to **Width** 5 and the circle next

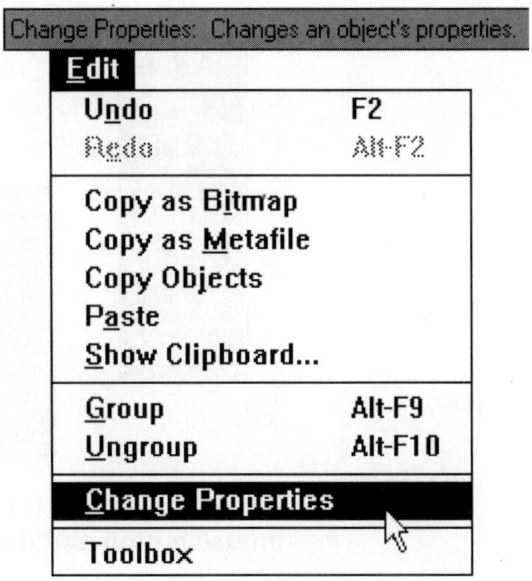

Fig. 5.2 The **Change Properties** tool from the **Edit** pull-down menu

to **Blank Fill** must be selected with a *left-click* in the dialogue box. See Fig. 5.3.

Similarly, to change the colour of an object, first select the required colour in the **Vertical Status Bar**, then select **Change Properties** from the **Edit** menu, and finally a *left-click* on the object to be changed results in the selected object's colour changing to the required colour.

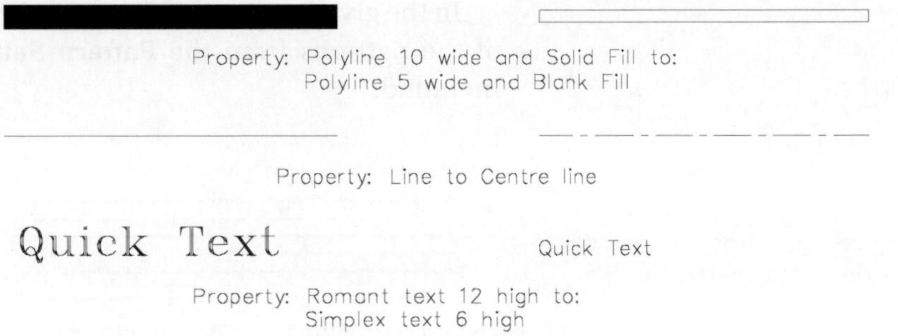

Fig. 5.3 Example 1

Example 2 – floating toolboxes

1. *Left-click* on the button in any toolbox. Fig. 5.4 shows the one in the **Draw** toolbox. *Left-click* on **Copy float** from the resulting pull-down menu.
2. *Left-click* on the arrow in the floating toolbox which appears.
3. Repeat 2 and the toolbox is rearranged.
4. Repeat 2 again, and so on – four possible changes in all.

   Fig. 5.5 shows that the floating toolbox can be altered in shape and dragged under mouse control to any position on screen. Position

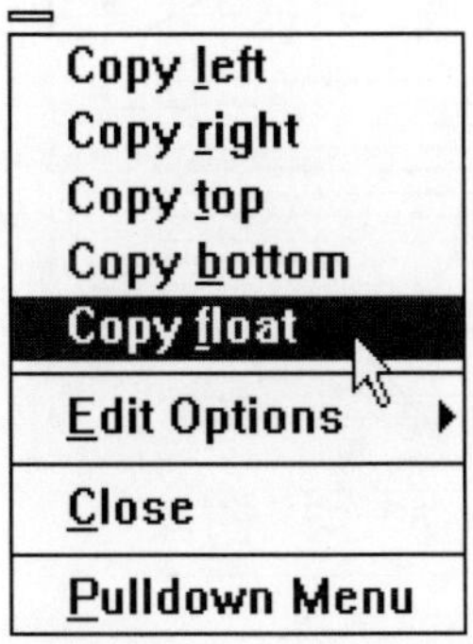

Fig. 5.4 The pull-down menu from a *left-click* on a toolbox button

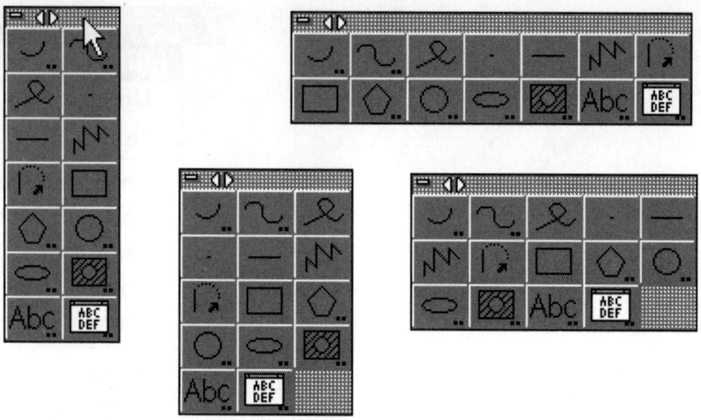

Fig. 5.5 Different forms of floating toolboxes

the cursor in the top bar of the floating toolbox, hold down the mouse button and drag.

Example 3 – Pattern Fill

With the **Pattern Fill** tool, many patterns can be included in drawings. Fig. 5.6 is an example of a simple building drawing – a two-view orthographic projection of a simple garden shed – which has had patterns added to the views with the aid of the **Pattern Fill** tool.

The **Pattern Settings** dialogue box has already been illustrated (page 21).

In the given orthographic projection, the names, angles and scales of the patterns from the **Pattern Settings** dialogue box have been included.

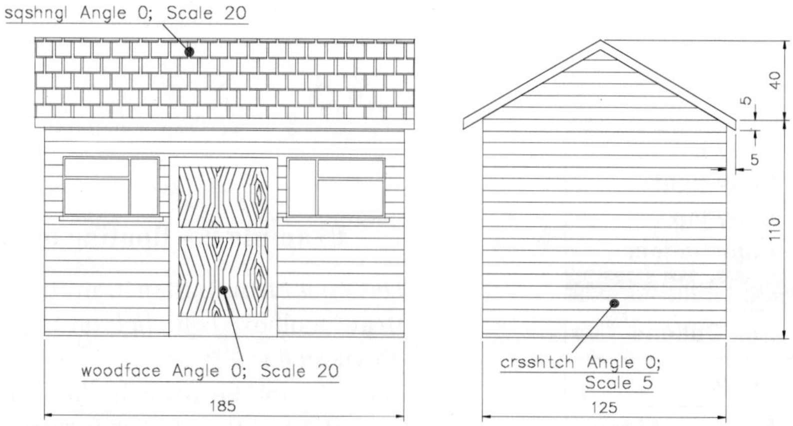

Fig. 5.6 Example 3

Example 4 – Pattern Fill: a second example

In this example, shown in Fig. 5.7, two **Polyline**s – an outer one 10 units wide and the inner 20 units wide – have been drawn, each with a pattern from the **Pattern Settings** dialogue box.

The central area outline was first drawn with the **Polyline** tool (width 0), then filled with a pattern from the **Pattern Settings** dialogue box. The three patterns used are:

The 10 unit wide polyline: **concrete** – Angle 0; Scale 1
The 20 unit wide polyline: **aggregat** – Angle 0; Scale 2
The central area pattern: **decotile** – Angle 0; Scale 1

Grid is set to 10; **Snap** is set to 5.

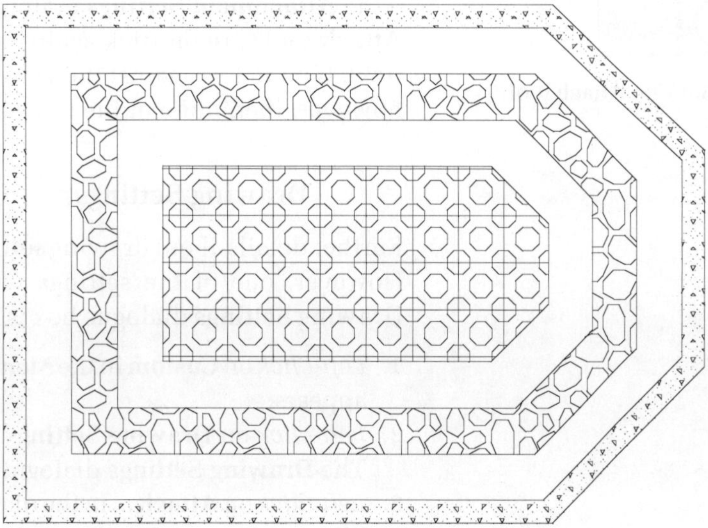

Fig. 5.7 Example 4

Example 5 – the Attach tools

The **Attach** tools have been mentioned in earlier pages. These tools are important in that, with their aid, new objects can be drawn with their start points and any intermediate points exactly established in relation to other objects already in a drawing. Each **Attach** tool, Fig. 5.8, has its own icon, with a 'parent' icon which highlights when any, some or all of the **Attach** tools are in operation. There are two methods of selecting an **Attach** tool:

1. Select the 'parent' icon and all **Attach** tool icons highlight, showing that they are currently operating. If necessary *left-click* on all icons other than those one wishes to be operative.

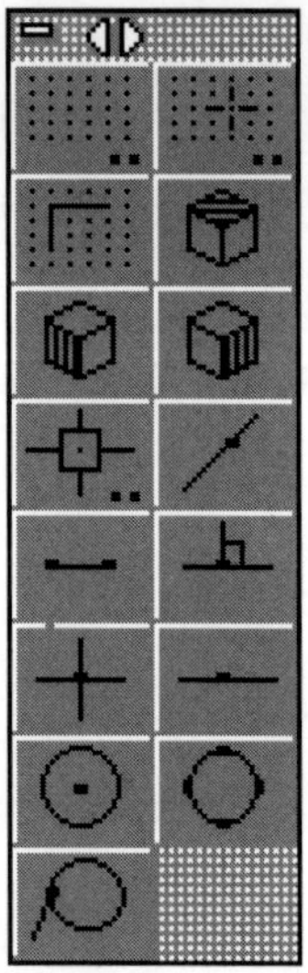

Fig. 5.8 The **Attach** tool icons

2. *Double-click* on any one of the icons. The 'parent' icon highlights and in addition only the selected icon also highlights.

Notes

One of the problems with having too many **Attach** tools operative at the same time is that if one selects an endpoint of an object and it happens to be near another attach point, the attachment may take place on the incorrect attach point. Personally I prefer a *double-click* on only those attach tools which I wish to be currently using.

Note that a *left-click* on the 'parent' attach icon sets *all* **Attach** tools – they will all highlight.

If you *double-click* on the 'parent' attach icon – it has a double square in its corner to show that it is associated with a dialogue box – the **Attachment Settings** dialogue box appears. This shows which **Attach** tools are on (tick against their name) and which are off (no tick). If no tick is against the word **Attach Mode** in the dialogue box, however, then none of the tools are currently available.

Drawing Settings

Further details about drawing settings will be given in a later chapter. However, one of the settings which can be made by calling the **Drawing Settings** dialogue box is the **Attach** tool settings – on or off.

1. *Left-click* on **Custom** in the **Menu bar**. The **Custom** pull-down menu appears.
2. *Left-click* on **Drawing Settings ...** in the **Custom** pull-down menu. The **Drawing Settings** dialogue box appears (Fig. 5.9).
3. *Left-click* on **Attach ...** in the dialogue box. The **Attachment Settings** dialogue box appears (Fig. 5.9).

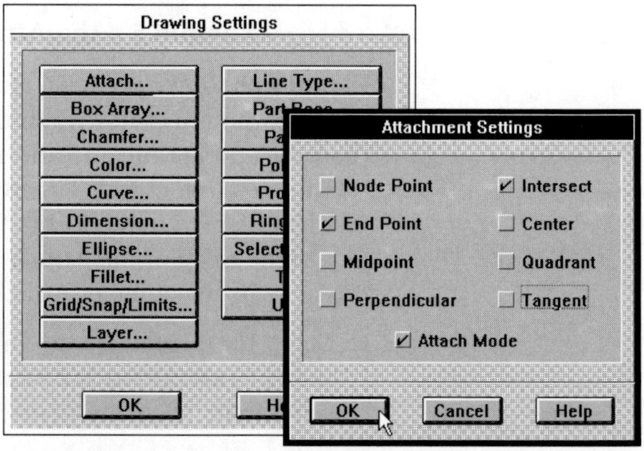

Fig. 5.9 The **Drawing Settings** dialogue box

4. Hold the mouse button down with the cursor over the top bar of the dialogue box and drag the dialogue box to a new position.
5. Set the required **Attach** tools with *left-clicks* in the **Attach Point** boxes next to the attach tool names.

Example 6 – AutoSketch Preferences

While dealing with part of the **Custom** menu, try the following:

1. *Left-click* on **Custom** in the **Menu bar**. The **Custom** pull-down menu appears.
2. *Left-click* on **Preferences ...** in the menu. The **AutoSketch Preferences** dialogue box appears (Fig. 5.10).
3. Experiment with **Background Colors** and **Toolbox Highlight Color**.

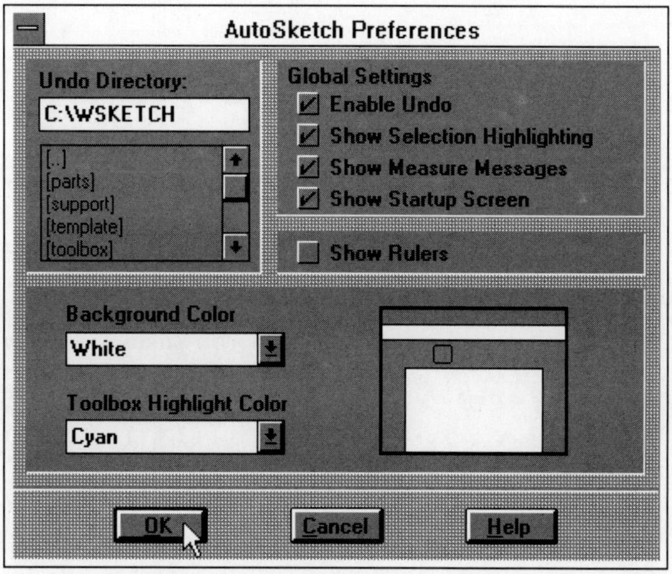

Fig. 5.10 The **AutoSketch Preferences** dialogue box

Exercises

Here are eleven exercises to help in revising the constructional methods given in earlier pages. These exercises involve the use of the tools from the **Draw** toolbox, from the **Assist** toolbox, and from the **Edit** toolbox. They are all intended to be worked on the already saved drawing template **A3SHEET**.

1. Fig. 5.11. A rear elevation of a building designed for student accommodation. All the windows are identical and have been drawn with **Snap** set to 2 units. Sizes do not really matter, the idea

of the exercise being to practise use of the **Rectangle** tool, the **Pattern Fill** tool and the **Box Array** tool.

Fig. 5.11 Exercise 1

2. Fig. 5.12. A **Box Array**. Start with a **Rectangle** 50 units square.

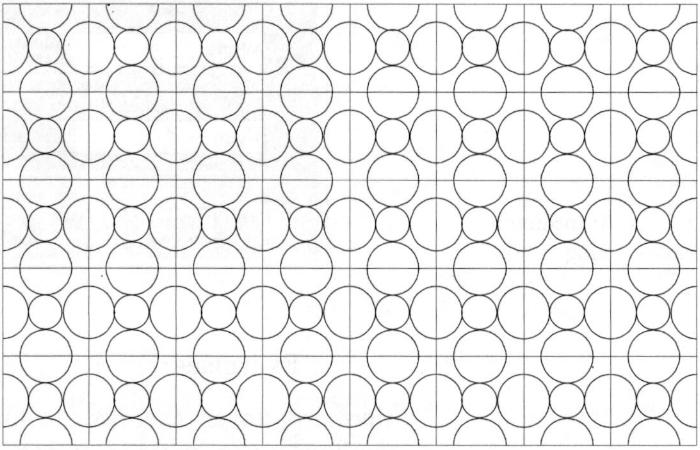

Fig. 5.12 Exercise 2

3. Fig. 5.13. An exercise to practise **Ring Array**. Details of the method of obtaining the part to be arrayed are indicated. If this drawing were intended for use in a machine shop, the sizes would be critical, but for this practice exercise, the diameter of the tiny circles does not matter a great deal.
4. Fig. 5.14. A pair of garden gates. **Snap** set to 2 units. Some detail results from the use of the **Box Array** tool.

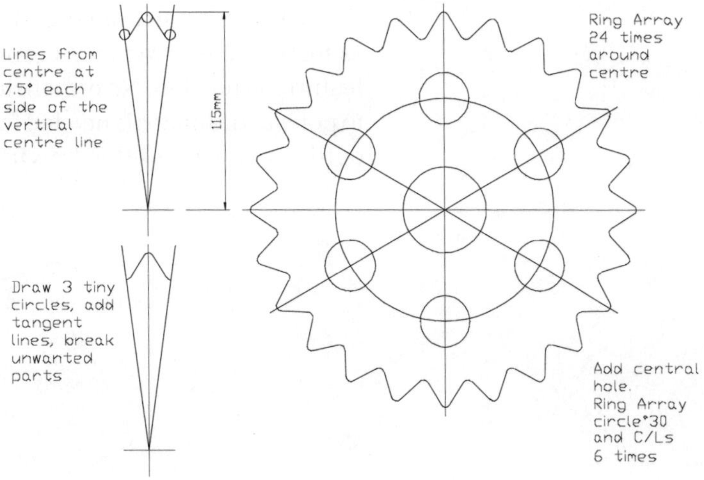

Fig. 5.13 Exercise 3

Fig. 5.14 Exercise 4

5. Fig. 5.15. AutoSketch is a valuable drawing tool for this type of 'poster' drawing.

Fig. 5.15 Exercise 5

6. Fig. 5.16. This site drawing should really be drawn with limits set as for an A1 or even an A0 drawing sheet. Careful positioning of features is needed. No dimensions are included, but drawing lines to polar coordinates is needed for obtaining accurate angles between outlines of parts of the drawing.

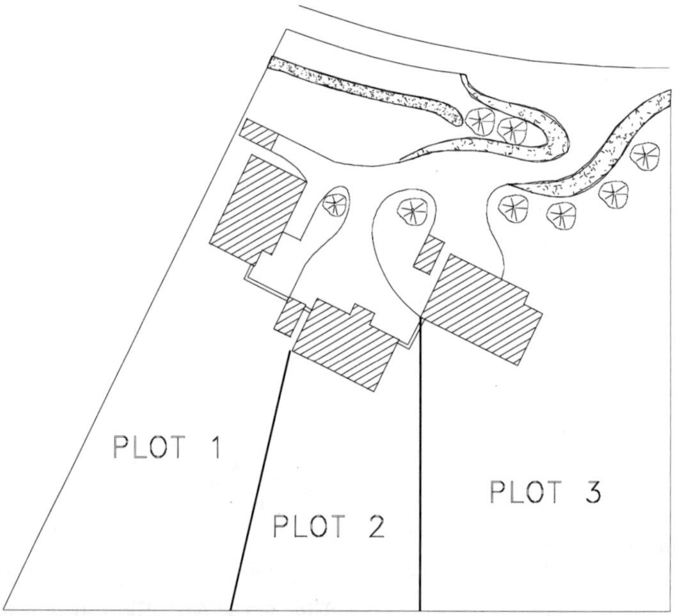

Fig. 5.16 Exercise 6

7. Fig. 5.17. A fairly detailed orthographic projection. **Snap** needs to be set to 2.5 units for parts of the front view outline. **Units** must be set to 1 digit after the decimal point when adding dimensions.

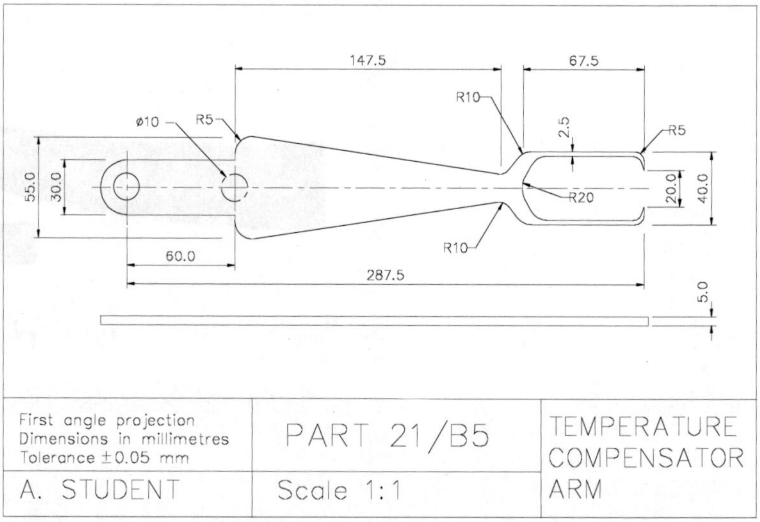

Fig. 5.17 Exercise 7

8. Fig. 5.18. Part of a ventilation system. The vents in the front view are drawn with the aid of the **Ring Array** tool. For some dimensions, set **Units** to 1 digit after the decimal point.

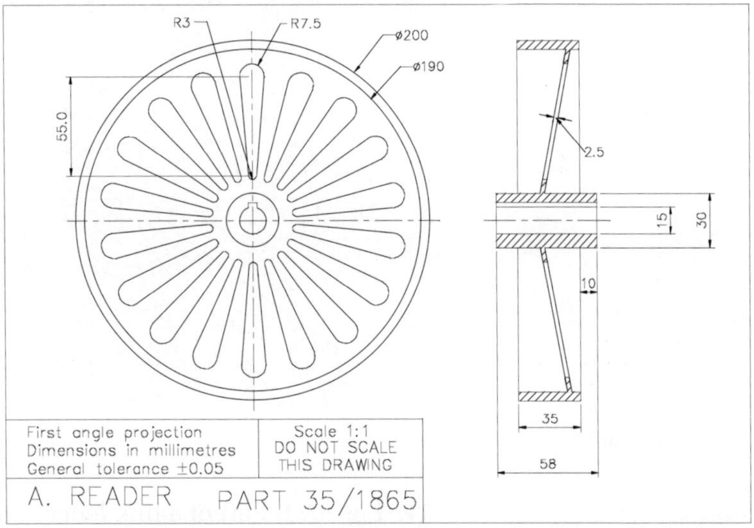

Fig. 5.18 Exercise 8

9. Fig. 5.19. A chart showing the never-ending nature of the design process. If the design does not satisfy the design brief, start all over again checking each feature of the process.

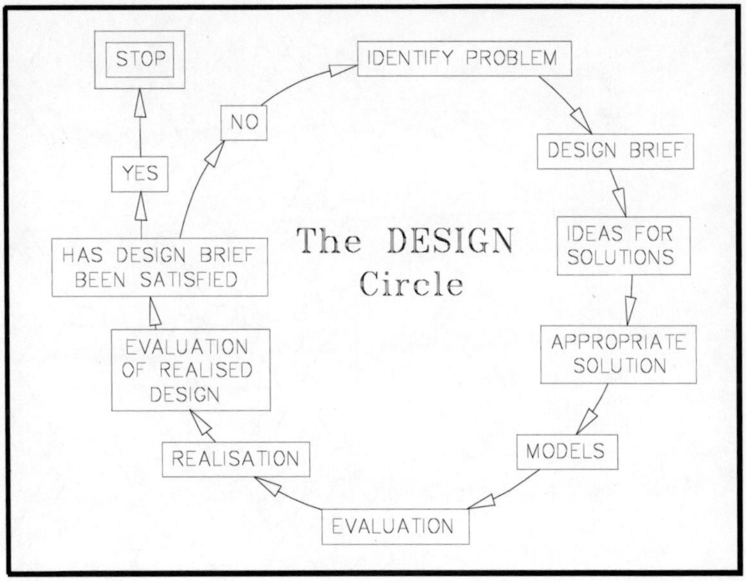

Fig. 5.19 Exercise 9

10. Fig. 5.20. Another **Ring Array**.

Fig. 5.20 Exercise 10

11. Fig. 5.21. Part of a link fence. Choose your own dimensions. The overlaps were achieved by constant use of the **Break** tool, with break points accurately selected by setting the **Intersect Attach** tool on.

Fig. 5.21 Exercise 11

CHAPTER 6

Parts and parts libraries

Introduction

Any AutoSketch drawing can be inserted into a drawing on screen with the aid of the **Part Clip...** tool from the **Edit** menu or toolbox (Fig. 6.1). This has particular relevance to inserting symbols of components into circuit drawings – electrical, electronic, pneumatic, plumbing – indeed into any circuit made up from standard symbols. The **Part Insert...** tool is for inserting **Parts** and **Clip Art** drawings into drawings on screen. One major use is to insert details such as nuts, bolts, washers and other fittings into engineering drawings. The operator can make up his own 'libraries' of symbols, or libraries of AutoSketch files (extension **.skd**) can be purchased ready drawn. If making up one's own symbol libraries, great care must be given to their construction because they may be used many times over. A large number of **Clip Art** files are included with the AutoSketch for Windows software package. **Parts** files are kept in the directory **wketch\parts** as sub-directories such as **wsketch\parts\electric**, **wsketch\parts\engineer**, **wsketch\parts\pneumat**, **wsketch\parts\clip_art** and so on. When the **Part** tool is called by a *left-click* on its icon (**Draw** toolbox), a dialogue box with a number of miniature drawings of the parts in the named directory appears on screen (Fig. 6.2). The required symbol is selected from this dialogue box.

Example 1 – constructing a symbol Part drawing file

1. Construct the drawing of the symbol – Fig. 6.3. Take care when constructing this drawing. It may be called back many times for circuit drawing.
2. *Left-click* on **Group** in the **Edit** menu.

   Group Select object: window the symbol

3. *Left-click* on **Part Clip ...** in the **File** pull-down menu. In the **Part Clip File** dialogue box: *left-click* on **[wsketch]** in the **Directories** list-box (Fig. 6.2).

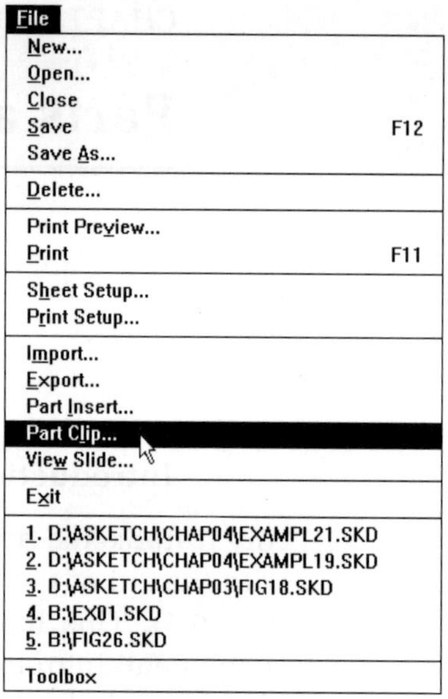

Fig. 6.1 **Part Clip...** in the **File** pull-down menu

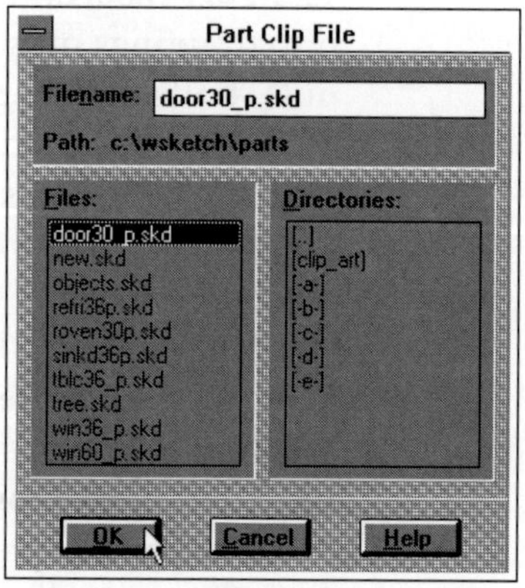

Fig. 6.2 The **Part Clip File** dialogue box

Double-click on **[wsketch]** in the **Directories** list-box
Double-click on **[parts]** in the **Directories** list-box
Double-click on **[pneumatic]** in the **Directories** list box

4. *Enter* 3-port02 in the **Filename** box. *Left-click* on **OK**.

   Part Clip Part base location: *left-click* on the required base point
   Part Clip Select object: *left-click* above and to left of symbol drawing

5. *Left-click* below and to right of drawing, and the part file is saved as file:

   c:\wsketch\parts\pneumati\3-port02.skd

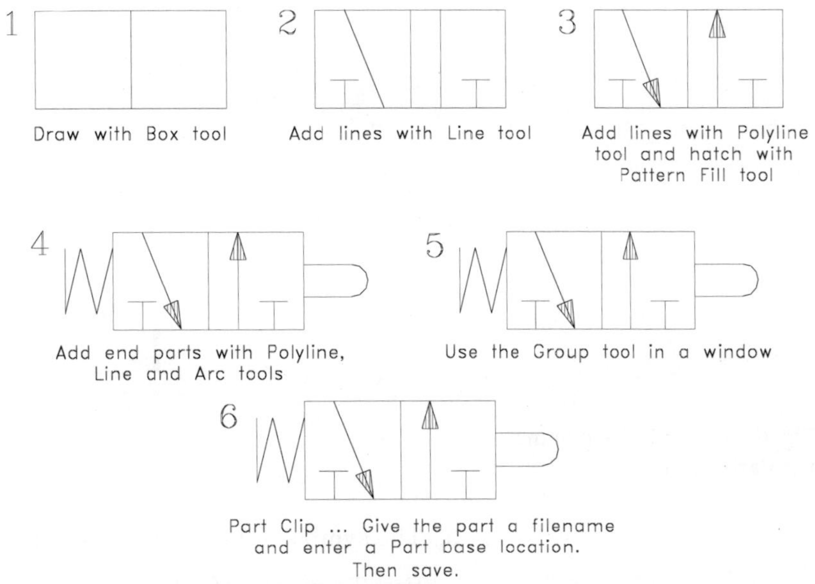

Fig. 6.3 Example 28

Example 2 – constructing a circuit drawing from Parts

1. *Double-click* on the **Part Insert ...** tool icon in the **File** toolbox.
2. The **Select Part File** dialogue box appears (Fig. 6.4).

   Double-click on **[-c-]** in **Directories**
   Double-click on **[wsketch]** in the **Directories** list box
   Double-click on **[parts]** in the **Directories** list box
   Double-click on **[electric]** in the **Directories** list box
   Left-click on the **npn** symbol. Check that the name in the **Active Filename** box is **npn.skd**. Check that there is a tick against **Rotate** in the **Insertion Option** box; *left-click* on **OK**

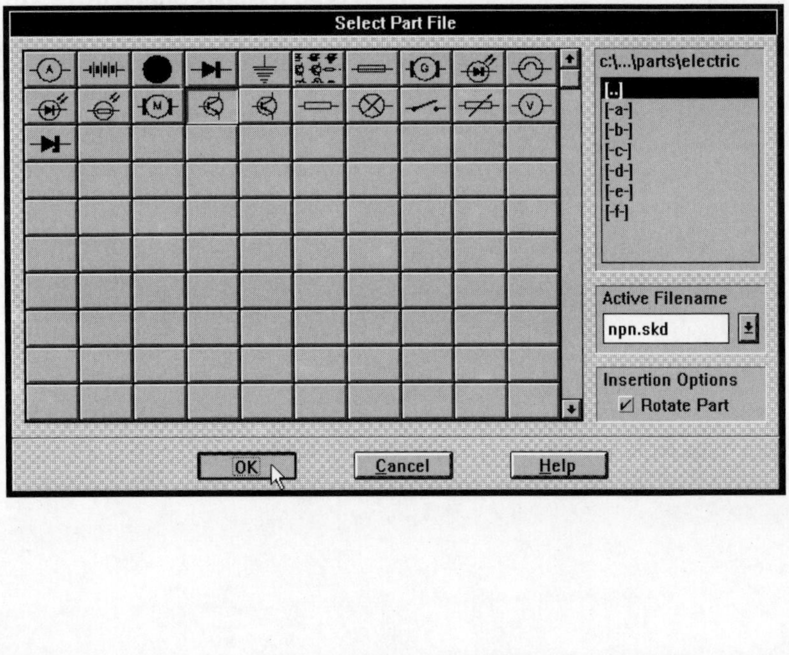

Fig. 6.4 The **Select Part File** dialogue box

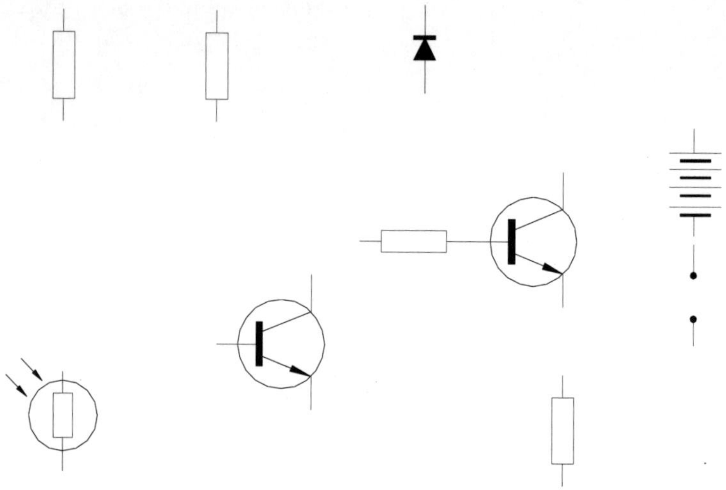

Fig. 6.5 Example 2 – placing the **Parts** symbols

3. The symbol appears on screen at the end of an arrow cursor.

   Part To point: *left-click*
   Part Rotation angle: (Angle: 0) *left-click* in this example
   Enter an angle figure for other symbols in the given circuit drawing
   Part To point: *left-click* at the point where the symbol is to be placed on screen

   Note: If a tick does not appear against **Rotate** in the **Insertion Options** box of the dialogue box, or if the tick is deleted by a *left-click* in the tick box when a tick is seen there, the symbol is placed on screen after only a single prompt – Part To point: appearing in the **Prompt** area of the AutoSketch drawing screen.
4. Insert other symbols from the parts sub-directory **c:\wsketch\parts\electric** in suitable positions on screen (Fig.

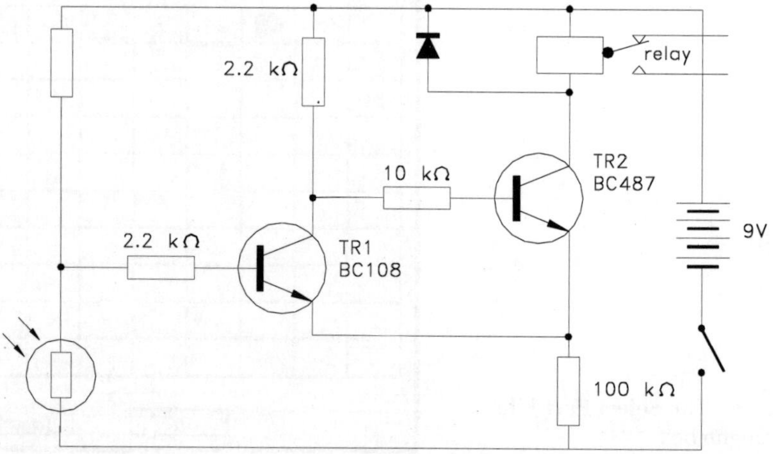

Fig. 6.6 Example 2 – the completed circuit drawing

6.5). For some of the symbols – the resistors, the diode, the battery, the lsr – an angle of rotation of 90 degrees will have to be entered.
5. When all symbols have been positioned, add connecting lines with the aid of the **Line** tool to complete the circuit (Fig. 6.6).
6. Construct symbols not included in the sub-directory in the circuit. In the example given, the relay was added in this manner.
7. Add connect dot symbols at conductor join points.
8. Add any necessary notes.

Examples of libraries

Three simple libraries of symbols, containing only a few symbols in each, are shown here:

1. The electronic circuit already given in Example 2 (Fig. 6.6), a Shmitt's Trigger circuit and the library of Parts symbols (Fig. 6.7) from which it was constructed.

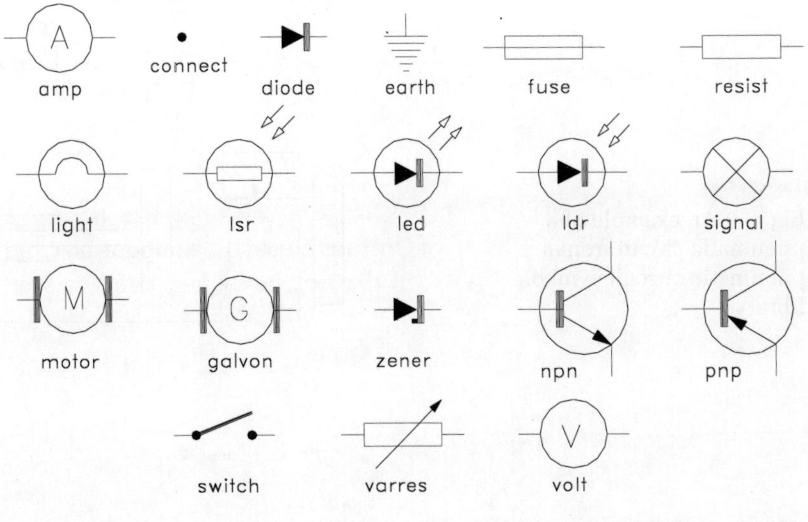

Fig. 6.7 A small library of electronic and electric circuit symbols

2. A pneumatics Parts symbols library (Fig. 6.8) together with a pneumatics circuit constructed from the symbols in the library (Fig. 6.9).
3. A building drawing Parts symbols library (Fig. 6.10), together with a plan of the first floor of a house, constructed partly from symbols in the library (Fig. 6.11).

Note: Libraries purchased from suppliers of disks of symbols may contain hundreds of symbols. The examples of libraries shown here are very small and are included only to indicate to the reader what a Parts library is. As an example of a commercially produced library, that from the AutoSketch for Windows **clip_art** sub-directory is a

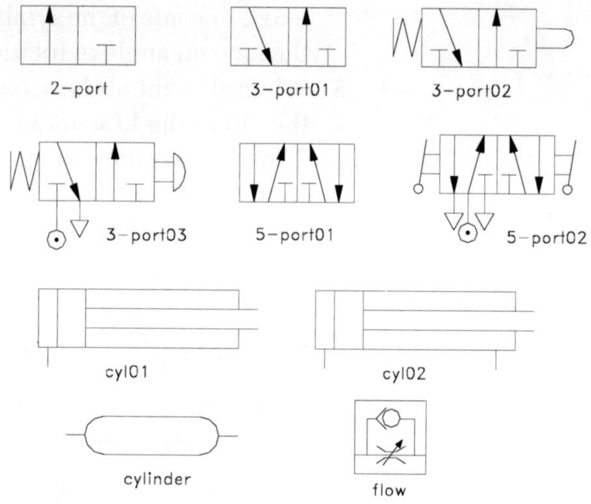

Fig. 6.8 A small library of pneumatic circuit symbols

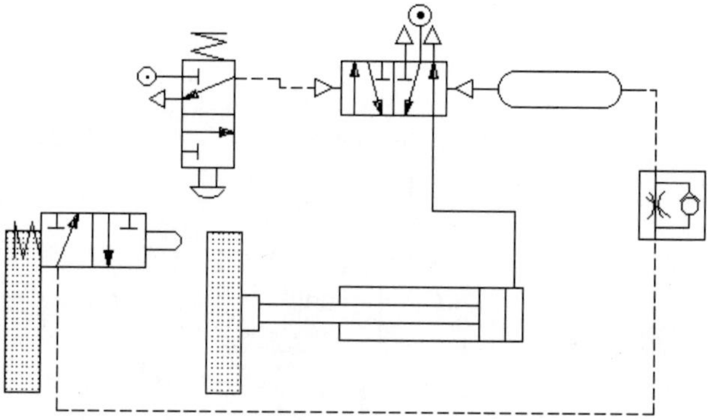

Fig. 6.9 An example of a pneumatic circuit from a pneumatic cirrcuit symbols library

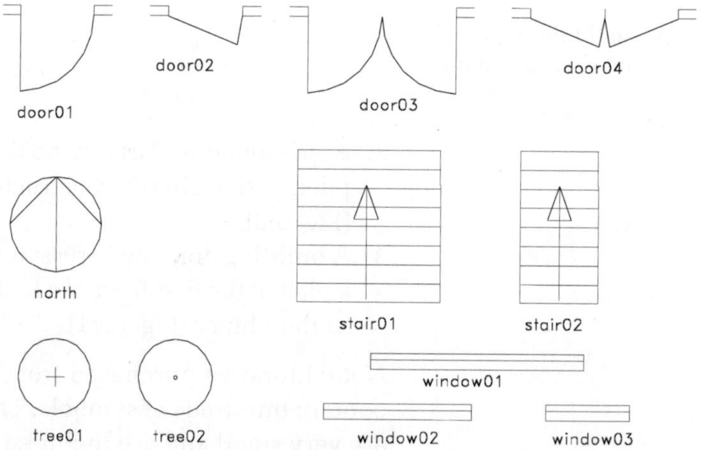

Fig. 6.10 A small library of building drawing symbols

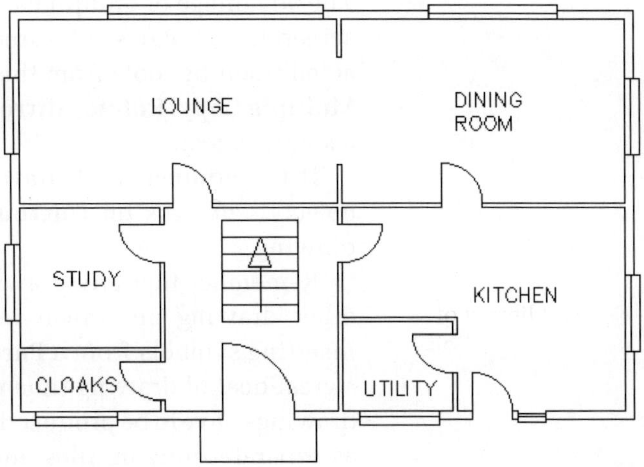

Fig. 6.11 A building plan drawn partly with symbols from a library

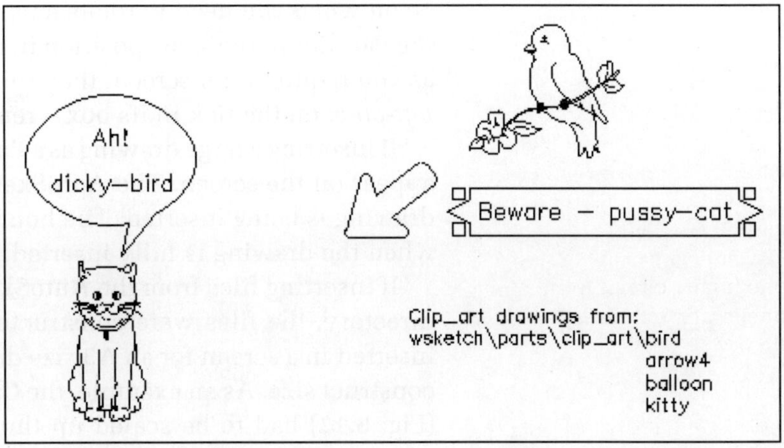

Fig. 6.12 An example of a drawing constructed from **Clip art** files

good one, with a large number of clip art drawing files. An example of a drawing containing clip art features is given in Fig. 6.12.

Notes on Part files

When constructing symbols, or any other type of drawing, as **Part** files, it is best to **Group** the objects making up the drawing. The method of doing this was described in Example 1 but is repeated here:

1. *Left-click* on the **Group** tool icon in the **Edit** menu.
2. Group Select objects: appears in the **Prompt** window: window the drawing – *left-click* top left, followed by a *left-click* bottom right of the drawing.

The advantage of grouping a **Parts** drawing is that when recalled to screen from a **Parts** sub-directory, the whole **Parts** drawing can be acted upon by tools from the **Modify** toolbox – **Move**, **Scale**, **Copy**, **Multiple Copy**, **Rotate**, **Mirror**, **Box Array**, **Ring Array** – as if it were a single object.

If the grouped part drawing is to be split up into its original objects, *left-click* on **Ungroup** in the **Edit** menu and window the drawing.

Remember that any drawing can be inserted as a **Part** into any other drawing on screen in exactly the same manner as when inserting symbols from a **Parts** sub-directory. This facility can save a great deal of drawing time when features common to a number of drawings have to be produced. Draw the common features, save them as separate drawing files and insert them as **Parts** in the various drawings being produced.

Note that the **Rotate** tick in the **Insertion Option** box should only be on when you need to rotate a part when it is placed on screen. If the part lies in the same position in the **Select Part File** dialogue box as you require it on screen, then it is quicker to turn **Rotate** off by a *left-click* on the tick in its box – removing the tick.

If inserting a large drawing as a **Part**, it may take a fair time to fully appear on the screen. The AutoSketch hourglass indicates that the drawing is being inserted. The hourglass reverts to an arrow cursor when the drawing is fully inserted.

If inserting files from the AutoSketch for Windows **Clip_art** sub-directory, the files were constructed for a 420,297 (A3) screen. If inserted in a screen for an A3 size drawing, they will appear at their construct size. As an example, the **Clip_art** drawings in the example (Fig. 6.12) had to be scaled up three times for them to appear on screen at a sensible size.

The drawing **order.skd** – a bonus drawing with the AutoSketch for Windows software – is an excellent example of a drawing constructed from a number of **Parts** file insertions. Remember, however, that it is a drawing designed for a 60 × 50 screen and if it is to be placed in an A3 drawing screen, it must be scaled up once it has been called.

Exercises

1. A diagram for a voltage dividing circuit is shown in Fig. 6.13. Taking the symbols in illustrations given earlier in this chapter:
   (a) Construct a sufficient number of electric/electronic symbols with which the circuit can be drawn as Parts files.

(b) By insertion of the Parts symbols you have drawn, construct the given circuit diagram.

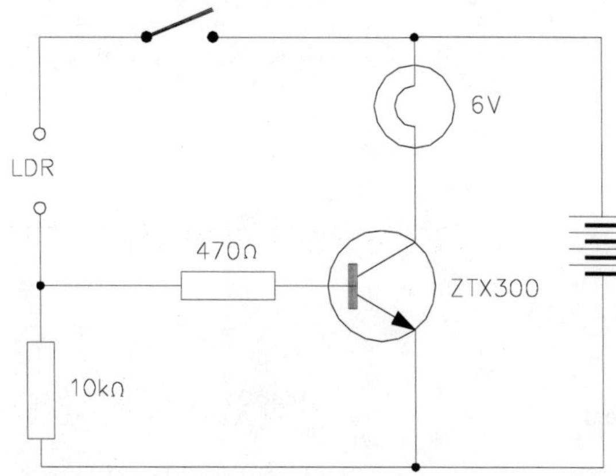

Fig. 6.13 Exercise 1

2. A simple electronic delay circuit is shown in Fig. 6.14. Copy the given drawing using the same methods as for Exercise 1. Add parts in the circuit which are not in your small symbols library when you have completed the Parts insertions.

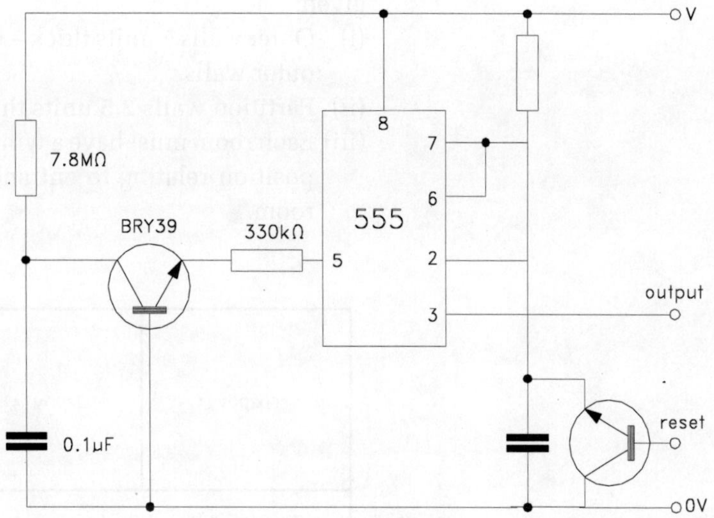

Fig. 6.14 Exercise 2

3. For the circuit diagram given in Fig. 6.15:
   (a) Construct the four pneumatics symbols required for this circuit – a vice closing/opening system – and save them as Parts files.
   (b) Construct the given circuit diagram (Fig. 6.15). Details not included with your Parts symbols should be added after they have been inserted.

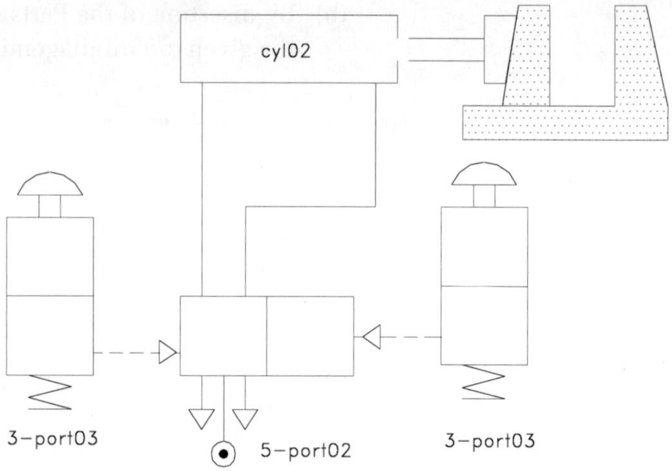

Fig. 6.15 Exercise 3

4. The building plan (Fig. 6.16) shows the upper storey of a two-storey four-bedroomed house. The dimensions included with the drawing are for your guidance only. With the building drawing symbols shown earlier in this chapter construct door and window symbols and add them to your library of Parts files. Working to the following details construct the required building plan to the dimensions given:
   (i) Outer walls 5 units thick – i.e. set **Snap** to 5 before drawing the outer walls.
   (ii) Partition walls 2.5 units thick – reset **Snap** to 2.5.
   (iii) Each room must have a window and a door placed in a sensible position relating to entrance to the room and lighting of the room.

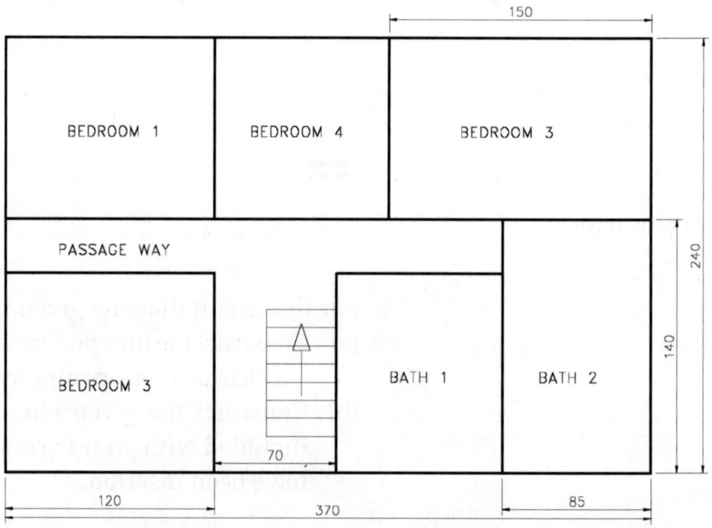

Fig. 6.16 Exercise 4

5. The drawing shown in Fig. 6.17, constructed from **clip_art** files given earlier in this chapter was drawn for the benefit of a five-year-old as part of a story she was being told. The given drawing is made up from four files from the sub-directory **c:\wsketch\parts\clip_art**. The four files are named:

   elephant.skd – drawn to two scales
   palm1. skd
   palm2.skd
   sun.skd

   Copy this, or design another drawing, using the **clip_art** files, which would also make a suitable illustration for part of a story for a five-year-old child.

Fig. 6.17 Exercise 5

CHAPTER 7

Pictorial drawing

Introduction

AutoSketch for Windows contains excellent facilities for constructing isometric drawings. In this chapter, methods of constructing isometric drawings with both axes at 30 degrees to the horizontal, oblique drawings with receding lines at 45 degrees and planometric drawings with both axes at 45 degrees, will be shown. Although the pictorial drawings shown here are all quite simple ones, AutoSketch can be used to construct quite complex drawings of this nature. It must be noted here that there is no way in which AutoSketch for Windows can be employed for constructing three-dimensional (3-D) drawings (solid models). The methods of pictorial drawing – isometric, oblique and planometric – are purely 2-D methods of drawing, even though they give the appearance on screen or when plotted or printed of being three-dimensional.

Isometric drawing

Fig. 7.1 shows examples of a number of simple isometric drawings constructed in AutoSketch for Windows. These drawings were constructed using the **Polyline** and **Ellipse** tools with the aid of the **Iso.Plane** tools. Fig. 7.2 shows the results of using the **Iso.Plane** tool icons. With the **A3SHEET.SKT** template opened:

1. Make sure the **Assist** toolbox is on the left hand side of the AutoSketch window.
2. *Left-click* on the **Grid** tool icon in the **Assist** toolbox.
3. *Left-click* on the **Snap** tool icon in the **Assist** toolbox.
4. *Left-click* on the **Left Iso.Plane** tool icon in the **Assist** toolbox. The grid dots arrange themselves in an isometric manner, sloping at 30 degrees to the horizontal in both directions. With the **Polyline** tool draw a 'square'.
5. *Left-click* on the **Top Iso.Plane** tool icon. With the **Polyline** tool draw a 'square'.

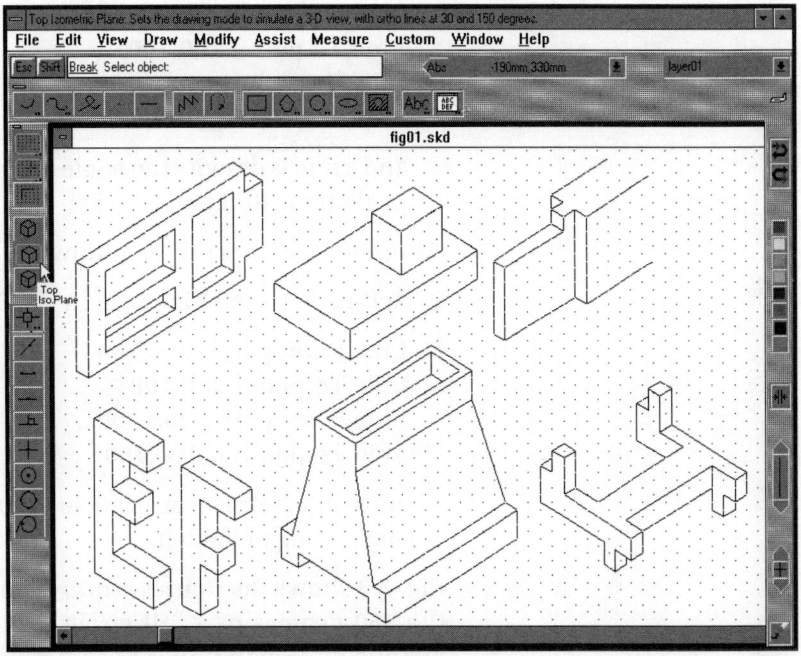

Fig. 7.1 Examples of isometric drawngs constructed in AutoSketch for Windows

6. *Left-click* on the **Right Iso.Plane** tool icon. With the **Polyline** tool draw a 'square'.
7. Now with the **Ellipse** tool, draw an ellipse with each of the **Iso.Plane** tools in action.

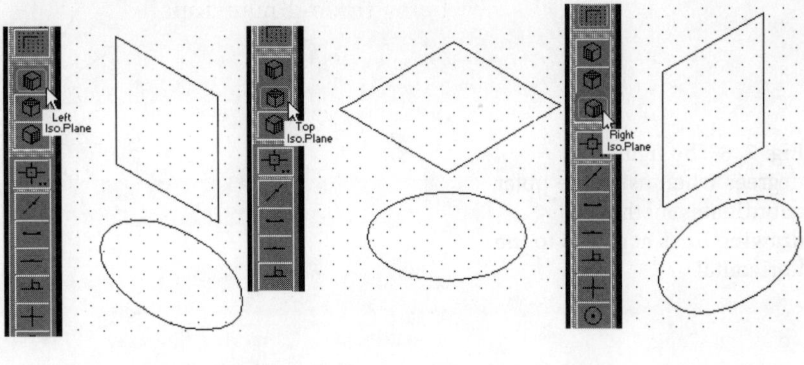

Fig. 7.2 The results of using **Polyline** and **Ellipse** with the **Iso.Plane** tools in action

Another method of constructing isometric drawings

Lines and polylines of required lengths can be drawn at angles to each other by using the polar co-ordinates method (see page 000). Two lines can be drawn at 30 degrees to each other as follows:

1. **New ...** file **A3SHEET.SKT**.
2. *Left-click* on the **Line** icon in the **Draw** toolbox.

Line From point: 100,100
Line To point: 200,100
Line From point: 100,100
Line To point: p(100,30)

and a line is drawn at an angle of 30 degrees to the first line, with the angle lying to the right.

Alternatively, proceed as follows:

2. *Left-click* on the **Line** icon in the **Draw** toolbox.

Line From point: 300,100
Line To point: 200,100
Line From point: 300,100
Line To point: p(100,150)

and a line is drawn at an angle of 30 degrees to the first line with the angle lying to the left.

Fig. 7.3 shows the figures required to draw lines at any angle within the 360 degrees of a full circle, when using the polar coordinates method of drawing lines at angles to the horizontal.

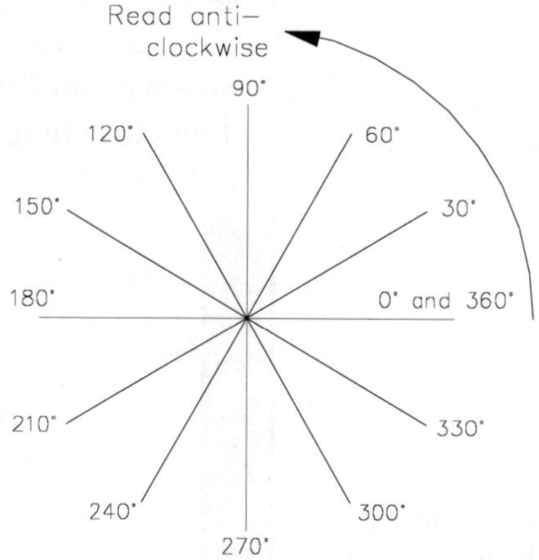

Fig. 7. 3 The figures for degrees when using the polar coordinates method of drawing lines at angles to the horizontal

Example 1 – constructing an isometric drawing

This example (Fig. 7.4) makes use of the polar coordinates method of drawing lines in order to construct an accurate isometric drawing which consists only of straight lines. Although this method is shown here, it is a time consuming method of construction when dealing with any drawing which is at all complicated. The reader is, however, advised to work this example – the method may occasionally be of value.

1. **New ...** A3SHEET.SKT.
2. *Left-click* on the **Polyline** icon in the **Draw** toolbox.

   Polyline From point: *left-click* at any point on screen
   Polyline To point: p(150,330)
   Polyline To point: p(100,30)
   Polyline To point: p(50,90)
   Polyline To point: p(150,150)
   Polyline To point: p(150,210)
   Polyline To point: p(50,270)

   This completes the first polyline, the top and part of the front face of the block being drawn.
3. *Left-click* on the **Polyline** icon in the **Draw** toolbox.

   Polyline From point: *left-click* at the left-hand **Re-start here** point shown in Fig. 7.4. Note that the **Assist Intersect** tool should be active
   Polyline To point: p(150,330)
   Polyline To point: p(50,270)

   This completes the front face of the block being drawn.
4. *Left-click* on the **Polyline** icon in the **Draw** toolbox;

   Polyline From point: *left-click* at the right-hand **Re-start here** point shown in Fig. 7.4. Note that the **Assist Intersect** tool should be active
   Polyline To point: p(100,30)

   Either *left-click* on the last point, or *left-click* on the **Polyline** icon in the **Draw** toolbox.

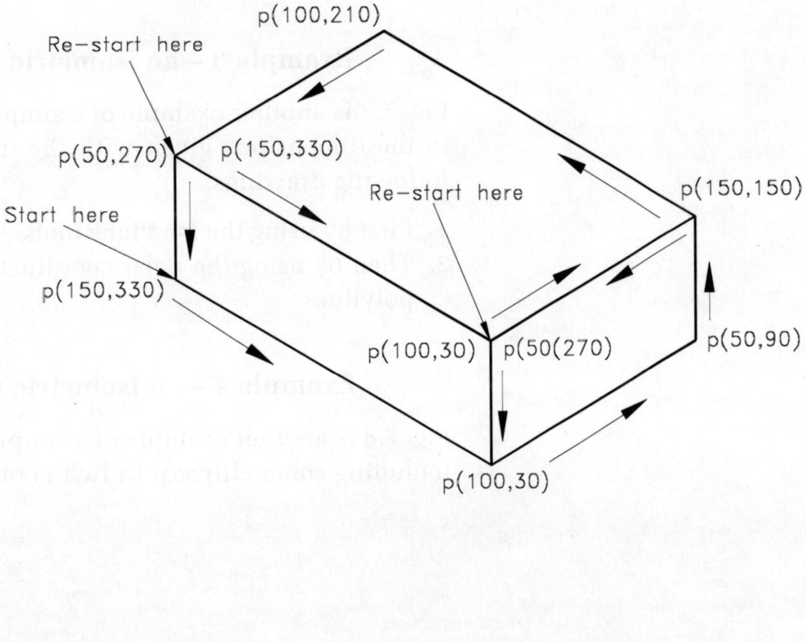

Fig. 7.4 Example 30

Note on ellipses in isometric drawings

Fig. 7.5 indicates the three angles at which ellipses lie on faces in isometric drawings. AutoSketch for Windows has an excellent ellipse drawing tool and ellipses will fit on to the isometric axes in action when the **Iso.Plane** tools are active as shown in Fig. 7.2.

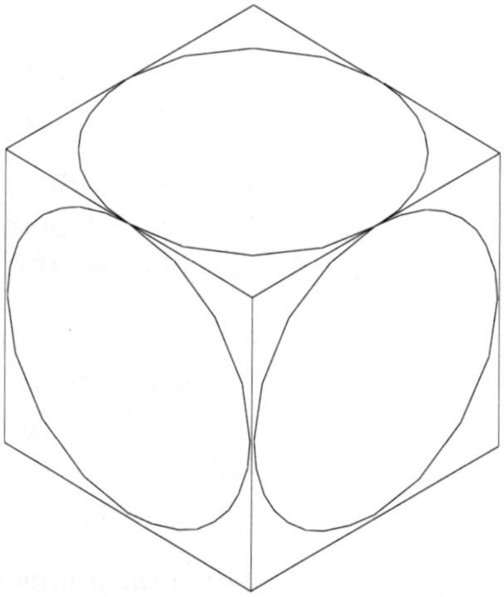

Fig. 7.5 Isometric ellipses on faces of an isometric block

Example 2 – an isometric drawing

Fig. 7.6 is an isometric drawing of a simple block with a centrally placed slot and with the two top edges chamfered. Construct this drawing:

1. First by using the **Iso.Plane** tools.
2. Then by using the polar coordinates method of drawing lines or polylines.

Example 3 – an isometric drawing

Fig. 7.7 is another example of a simple isometric drawing. Working to the dimensions given with the drawing, construct both of the following drawings:

1. First by using the **Iso.Plane** tools.
2. Then by using the polar coordinates method of drawing lines or polylines.

Example 4 – an isometric drawing

Fig. 7.8 is another example of a simple isometric drawing, this time including some ellipses. Draw the outer ellipse, then copy it to its

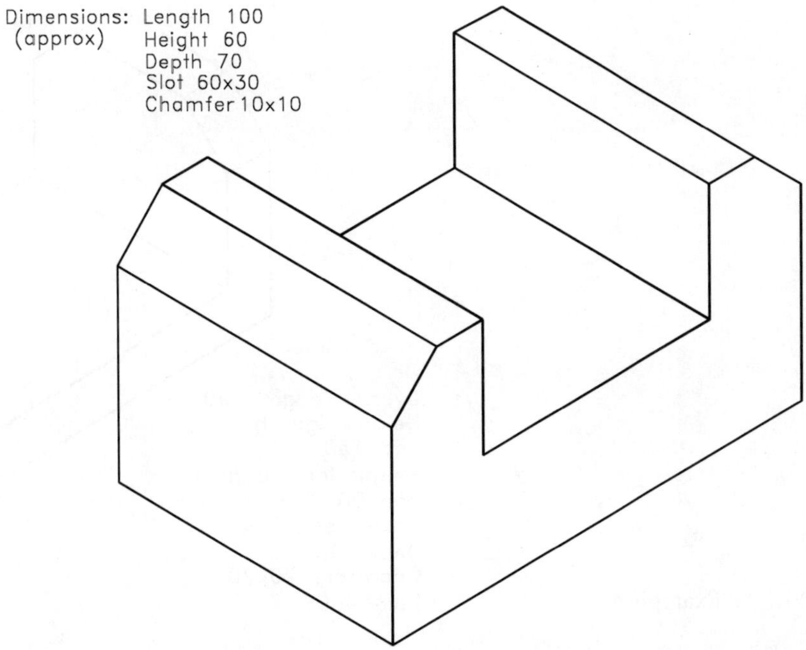

Fig. 7.6 Example 2

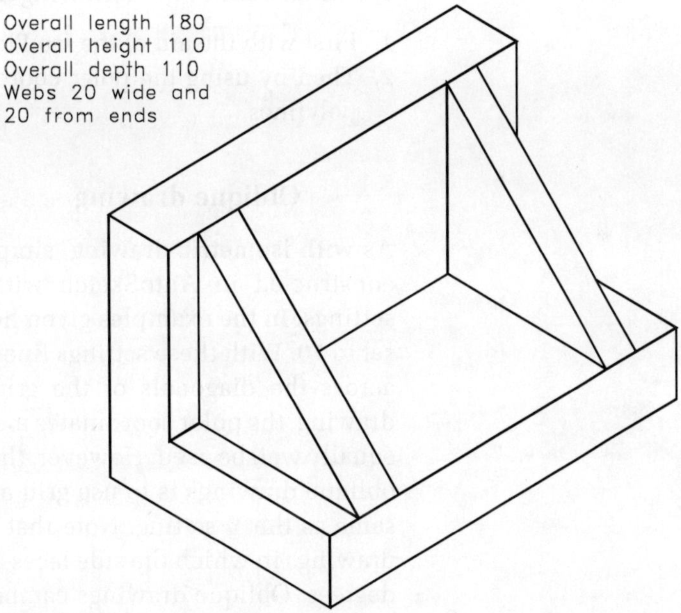

Fig. 7.7 Example 3

inner position – the grid points and snap will assist here. Then use the **Break** tool from the **Edit** toolbox to break off that part of the ellipse not required. Note that the **Intersect** tool from the **Assist** toolbox needs to be on to facilitate the breaking at the exact required

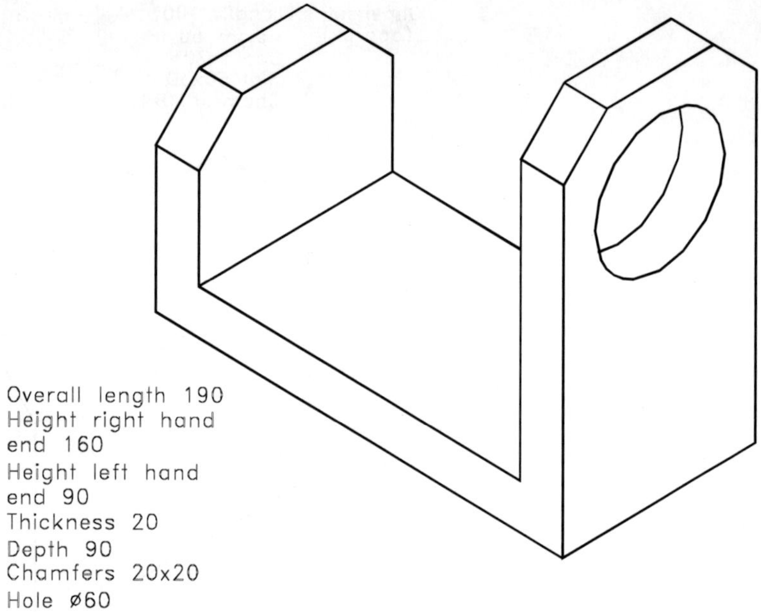

Fig. 7.8 Example 4

positions. Working to the dimensions given with the drawing, construct both of the following drawings:

1. First with the aid of the **Iso.Plane** tools.
2. Then by using the polar coordinates method of drawing lines or polylines.

Oblique drawing

As with isometric drawing, simple oblique drawings can easily be constructed in AutoSketch with **Grid** and **Snap** to appropriate settings. In the examples given here, **Snap** is to be set to 5 and **Grid** set to 10. With these settings lines at 45 degrees can easily be drawn across the diagonals of the grid dot pattern. As with isometric drawing, the polar coordinates method of drawing lines at angles can equally well be used. However, the easiest constructional method for oblique drawings is to use grid and snap set with the X setting the same as the Y setting. Note that we are only dealing with oblique drawings in which the side faces are drawn with receding lines at 45 degrees. Oblique drawings can be drawn with the receding lines at any appropriate angles when working ‘by hand’. Note also that, if required, oblique drawings can be drawn as cabinet drawings, in which receding lines along the 45 degrees axis are generally drawn at half scale.

Fig. 7.9 shows a number of examples of simple oblique drawings constructed in an AutoSketch screen with **Grid** settings of **X:** = 10 and **Y:** = 10 and **Snap** settings of **X:** = 5 and **Y:** = 5. In general, oblique drawing as a pictorial method tends to show distortions. Such drawings do, however, have the advantage that if one side of the object being drawn contains arcs, circles or curves, the drawings are particularly easy to construct. However, if the arcs, circles or curves occur on more than one face, the constructions can prove to be much more difficult.

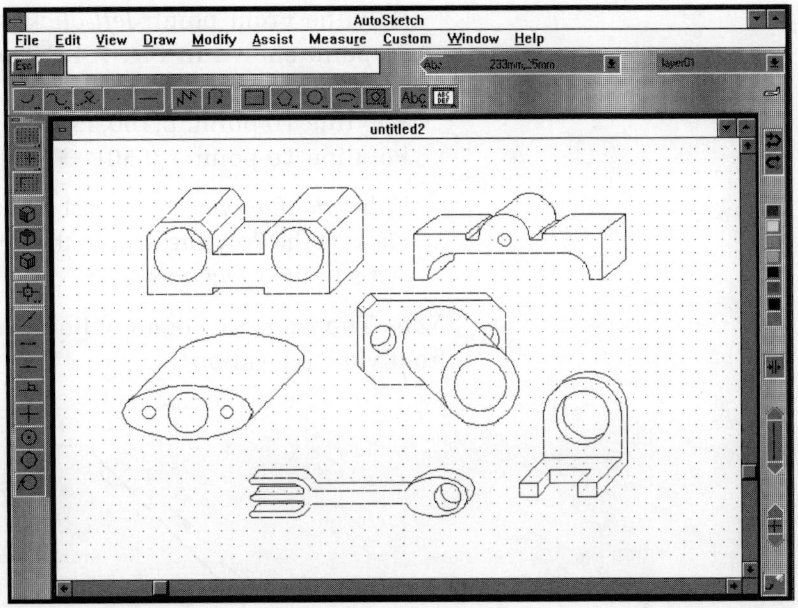

Fig. 7.9 Examples of oblique drawing constructed on a grid pattern with X and Y settings equal

Example 5 – constructing an oblique drawing

1. **New... A3SHEET.SKT**.
2. *Left-click* on the **Polyline** icon in the **Draw** toolbox.

   Polyline From point: *left-click* at any point on screen
   Polyline To point: r(150,0)
   Polyline To point: r(0,50)
   Polyline To point: r(-150,0)
   Polyline To point: r(0,-50)

   This completes the first polyline, the front face of the block being drawn.
3. *Left-click* on the **Polyline** icon in the **Draw** toolbox.

   Polyline From point: *left-click* at the left-hand **Re-start here**

point shown in Fig. 7.4. Note that the **Assist Intersect** tool should be active
Polyline To point: p(100,45)
Polyline To point: r(150,0)
Polyline To point: p(100,225)

This completes the top face of the block being drawn. Either *double-left-click* at the last point reached by the polyline, or *left- click* on the **Polyline** icon in the **Draw** toolbox.

4. *Left-click* on the **Polyline** icon in the **Draw** toolbox.

Polyline From point: *left-click* at the right hand **Re-start here** point shown in Fig. 7.4. Note that the **Assist Intersect** tool should be active
Polyline To point: p(100,45)
Polyline To point: r(0,50)

Either *double-click* on the last point reached by the polyline, or *left-click* on the **Polyline** icon in the **Draw** toolbox. Fig. 7.10 shows the steps in the procedure.

Note: The front face of an oblique drawing is often the same as its front view (elevation).

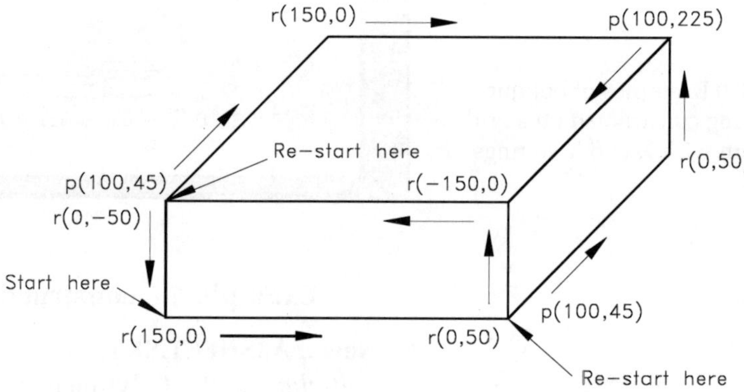

Fig. 7.10 Example 34

Example 6 – an oblique drawing

Fig. 7.11 – a child's walking trolley – is an example of an oblique drawing which requires some care to construct. Working to any suitable dimensions, construct the drawing by using the method of setting up your **A3SHEET** template with **Snap** and **Grid** set to **X:** = 10, **Y:** = 10. Note that the construction of this example should be in four stages:

1. Construct the wheels at the front.

2. Construct the front vertical handle support – it must be set back by the wheel thickness along the 45 degrees axis, i.e. across the diagonals of the grid dot pattern.
3. Construct the box, which is again set back from the front of the handle support.
4. Construct the rear handle support, which is again set to the back of the box. Then add the handle itself.

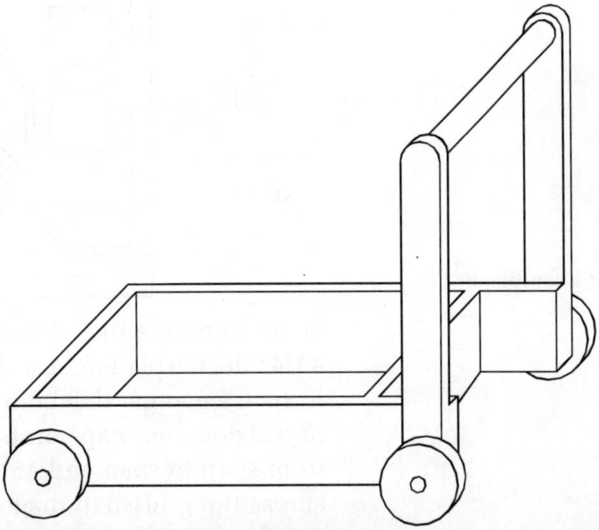

Fig. 7.11 Example 6

Example 7 – an oblique drawing

Fig. 7.12 – a three-socket electric plug – is another example of an oblique drawing, which requires some care to construct. Working to any suitable dimensions, construct the drawing by using the method of setting up your **A3SHEET** template with **Snap** and **Grid** set to **X:** = 10, **Y:** = 10. Note that the construction of this example should be in four stages:

1. Construct the front of the plug, from which the pins project.
2. Construct the projecting pins.
3. Construct the box, which is set back from the front face already drawn.
4. Construct the slots in the top and side.

Planometric drawing

Planometric drawing based on a plan of the item or article being drawn, which is tilted at 45 degrees, is a useful method of producing pictorial drawings of features such as houses, room plans, garden

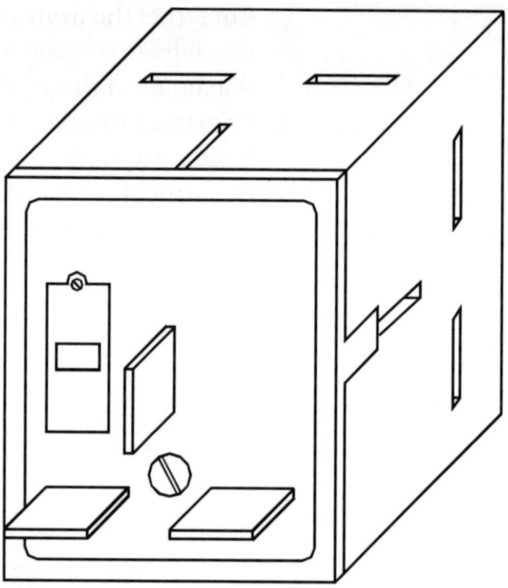

Fig. 7.12 Example 36

layouts and the like. AutoSketch lends itself to the construction of 45/45 degree planometric drawings, because with **Grid** and **Snap** set in such a manner that both **X:** and **Y:** figures are the same, a network of grid dots and snap points is set up so that verticals are easily drawn from snap to snap and 45 degree lines are equally as easily drawn across the grid/snap diagonals. Fig. 7.13 is an example of a 45/45 degree planometric drawing of a bungalow with its garage shown in the AutoSketch screen.

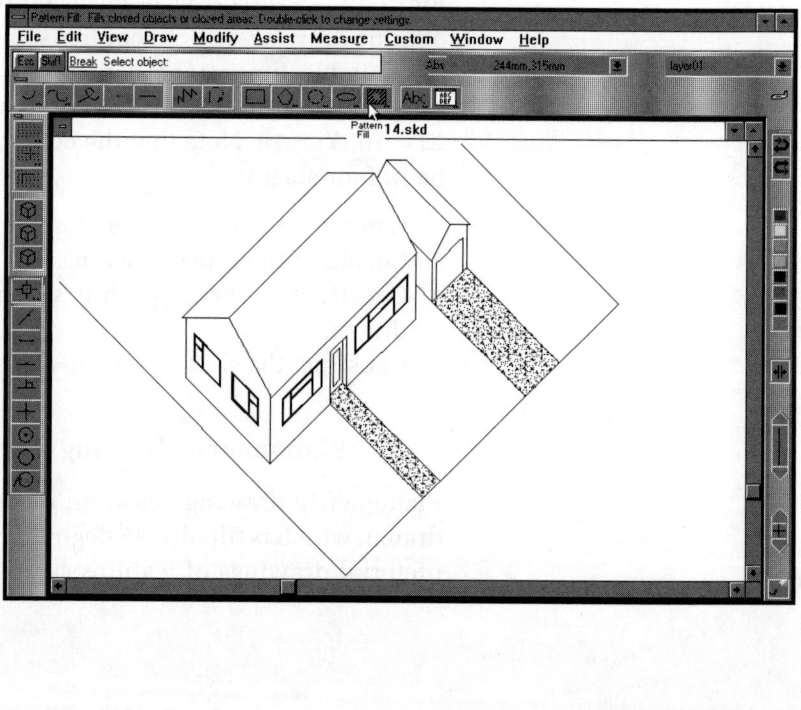

Fig. 7.13 Planometric drawing in the AutoSketch screen

Fig. 7.14 is a plot of the same drawing made from the screen drawing. Parts of the drawing have been drawn with the aid of the **Line** tool. Other parts have been drawn with the aid of the **Polyline** tool set to 1 unit thickness. Note the addition of the concrete **Pattern Fill** on the paths. Other pattern fills can be added to this form of drawing providing it is remembered that the pattern angles may not fit in the shapes of features such as walls, roofs, etc.

Example 8 – a planometric drawing

Working to dimensions of your own choice construct a planometric drawing of a bungalow similar to that shown in Fig. 7.14.

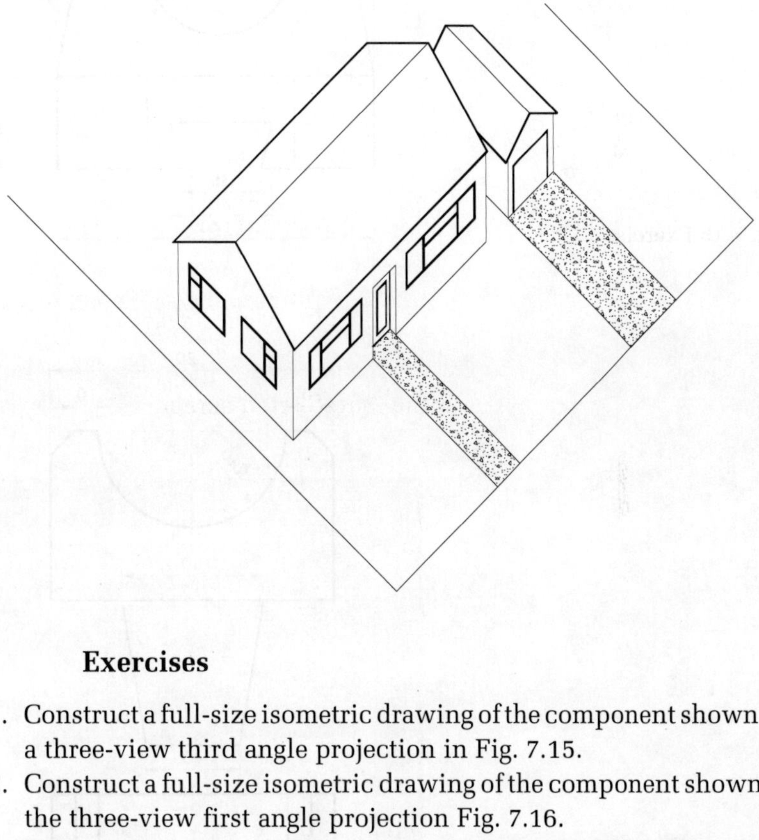

Fig. 7.14 Plot of Example 8

Exercises

1. Construct a full-size isometric drawing of the component shown by a three-view third angle projection in Fig. 7.15.
2. Construct a full-size isometric drawing of the component shown in the three-view first angle projection Fig. 7.16.
3. Taking any face as its front, construct an oblique drawing of the electric cable connector shown in the two-view third angle projection of Fig. 7.17.
4. Construct an oblique drawing of the car seat connector shown in the first angle projection of Fig. 7.18.

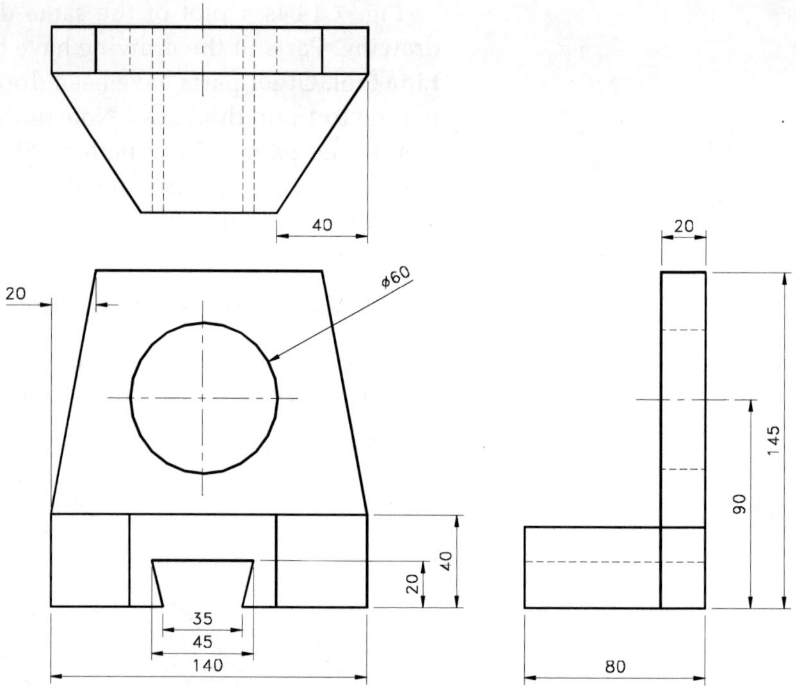

Fig. 7.15 Exercise 1

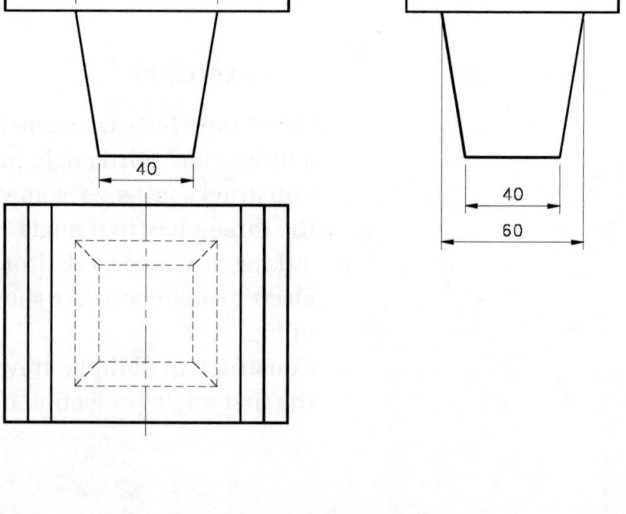

Fig. 7.16 Exercise 2

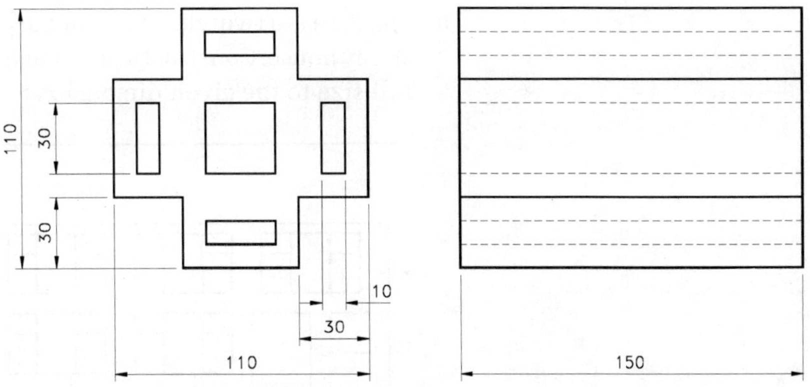

Fig. 7.17 Exercise 3

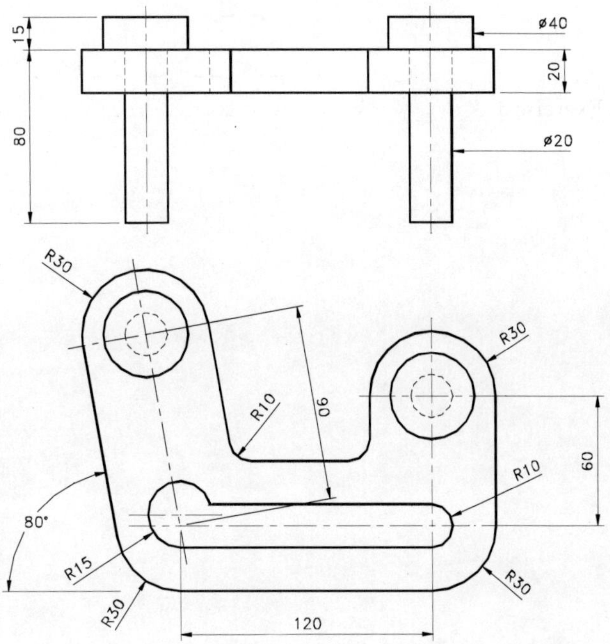

Fig. 7.18 Exercise 4

5. Fig. 7.19 is a two-view third angle projection of the outline of a two-storey house. Construct a planometric drawing of the house, working full size to the given dimensions.

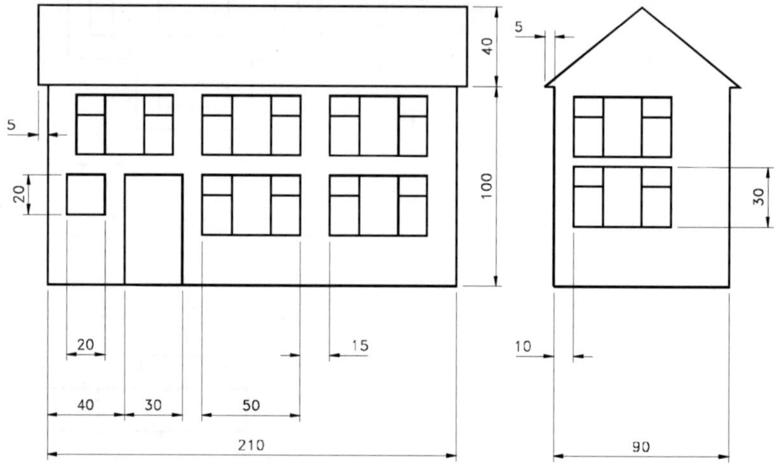

Fig. 7.19 Exercise 5

CHAPTER 8

The Measure toolbox

Introduction

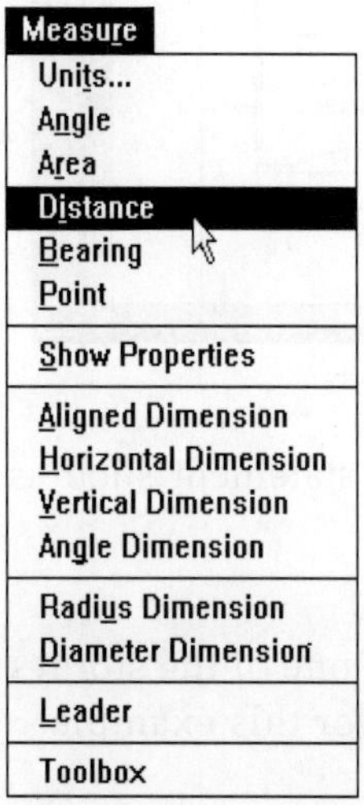

Fig. 8.1 The **Measure** pull-down menu

As with other tools in AutoSketch for Windows, the Measure tools can be selected either from a pull-down menu or by selection from tool icons in a toolbox. Fig. 8.1 shows the **Measure** pull-down menu which appears on screen after a *left-click* on **Measure** in the **Menu bar**. Fig. 8.2 shows the **Measure** toolbox with its icons along the top edge of the AutoSketch screen.

Apart from tools which can show measurements of Angle, Area, Distance, Bearing and Point position, various forms of dimensioning are controlled by tools from **Measure**. The only other tool in this toolbox is **Show Properties** – for displaying in a dialogue box on screen the properties of any objects selected from those within a drawing on screen.

Note: In the examples in this chapter, a different computer and monitor were used for producing illustrations. In addition the Windows colours setup was changed from that used previously. When any detail in the Windows setup is changed, it will take effect in AutoSketch for Windows.

Example 1 – the Distance tool

Open any drawing you may have saved to disk. The example shows plans of the layouts of the two storeys in a house, but any drawing will be suitable.

1. *Left-click* on the **Intersect** tool in the **Assist** toolbox. This will ensure that points chosen when measuring from points with the aid of the **Measure** tools are selected with accuracy.
2. *Left-click* on the **Measure Distance** tool in the **Measure** toolbox.

   Distance From point: *left-click* at any position in your drawing
   Distance To point: *left-click* at another point

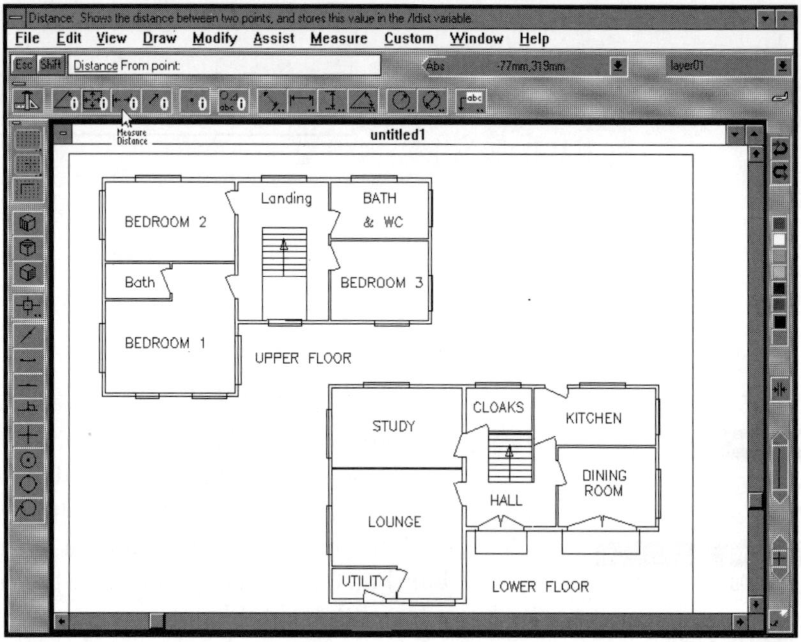

Fig. 8.2 The **Measure** toolkit along the top edge of the AutoSketch screen

The **Distance** information box appears with a statement such as **Distance is 180mm.**

3. *Left-click* on **OK** in the information box.

Fig. 8.3 is an example of the length of one side of one of the storeys in the house plans chosen as a suitable drawing for this example.

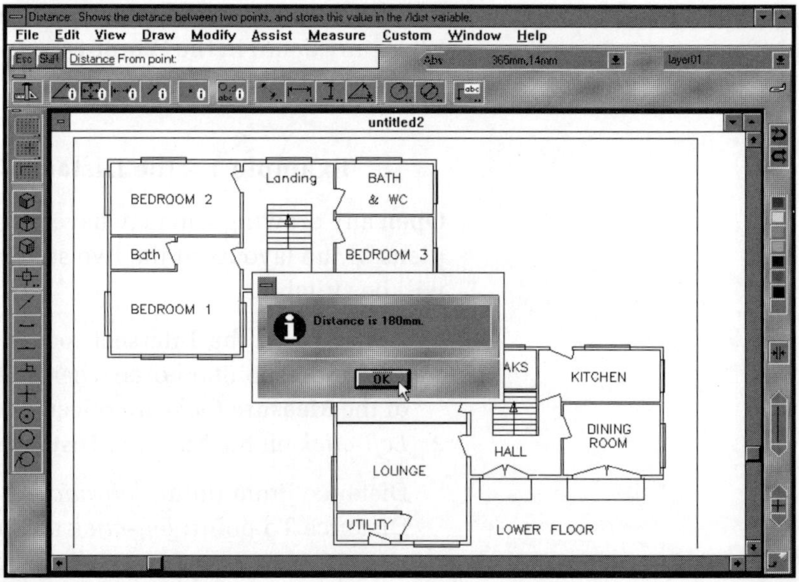

Fig. 8.3 An example of a **Distance** information box

Note: Distances and areas measured with the **Measure** tools in this example are in millimetre units. The actual scaled distances depend upon the scale to which the drawing has been constructed. In the example given, if the plan had been drawn to a scale of 1:50, the reading of 180 mm units would be translated to give the actual length of the wall as 9 metres – at a scale of 1:50, each millimetre represents 50 mm. Thus 180 multiplied by 50 = 9000 mm = 9 metres.

Example 2 – the Area tool

Left-click on the **Measure Area** tool in the **Measure** toolbox.

Area First perimeter point: *left-click* on the first corner of that part of the drawing whose area is to be found
Area Next point: *left-click* on the next corner of the area
Area Next point: *left-click* on the next corner of the area

And so on around the perimeter of the area. When the first perimeter point is selected, a cross at 45 degrees appears at the point. As each next point is selected further crosses appear at the points. The first perimeter point cross is larger than the others. When the first perimeter point is re-selected, an information box appears with details of the area and perimeter in coordinate units – Fig. 8.4.

Note: As with distance, so with area – the figures appearing in the Area information box are in coordinate units. Remembering that a square metre contains 1,000,000 square millimetres, if working to a

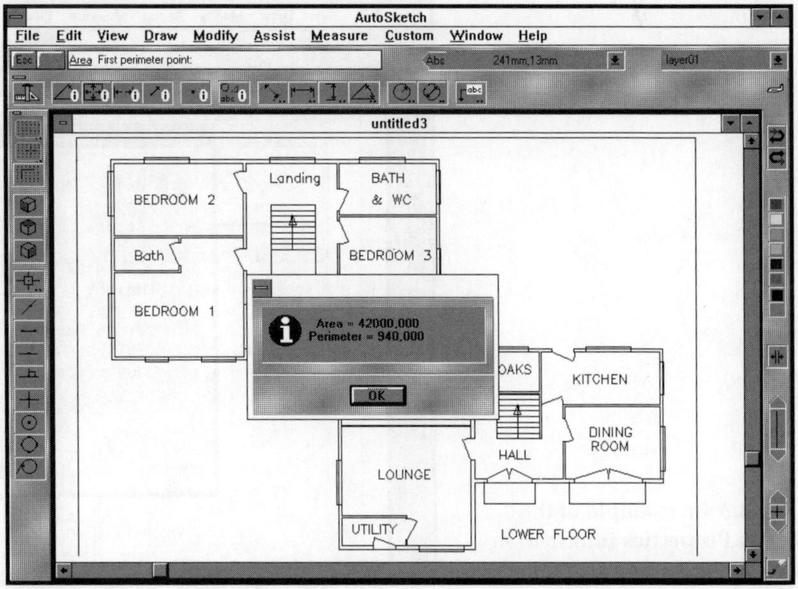

Fig. 8.4 The **Area** information box

scale of 1:50 each square metre is represented on the drawing by a square of sides 20 mm. Thus 1 square metre is 20 ¥ 20 = 400 square mm on the drawing. Thus the reading of 42000.000 shown in Fig. 8.4 represents 42000 divided by 400 =105 square metres.

Example 3 – Show Properties

1. *Left-click* on the **Show Properties** icon in the **Measure** toolbox. A hand-like pointer appears under mouse control. Show Properties Select object: *left-click* on the object whose properties are required. In the example in Fig. 8.5, an item of text was selected. The **Show Properties** information box appears on screen, detailing the properties of the selected object. A complete list of the properties of the chosen object is given in the information box.
2. Repeat 1 selecting various types of objects from your drawing, e.g. lines, circles, arcs, etc.

Dimensions

Some details about the simple horizontal, vertical and leader dimensioning were given in an earlier chapter. Further details are given below. The tools for adding dimensions to drawings are contained in the **Measure** toolbox and, as can be seen by reference to Fig. 8.1, these tools allow Aligned dimensions, Horizontal dimensions, Vertical dimensions, Angle dimensions and Leaders to be included in a drawing. A variety

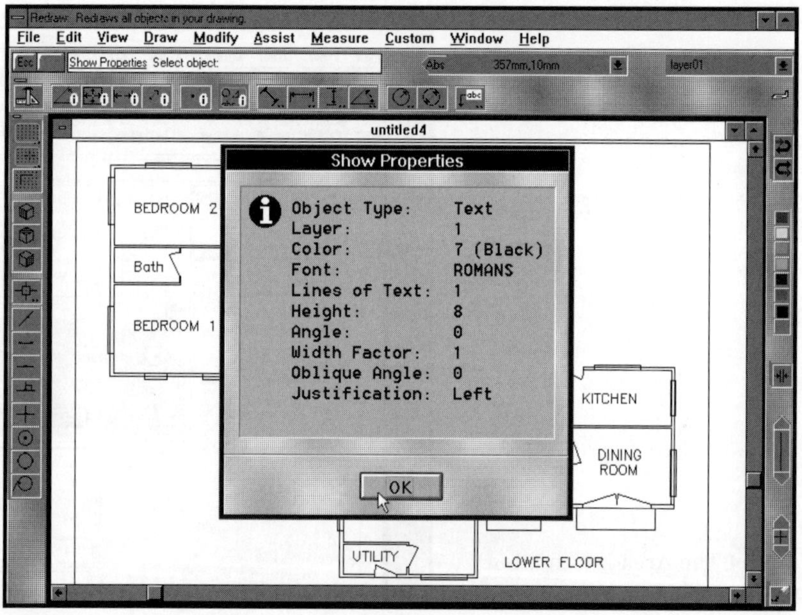

Fig. 8.5 An example of the **Show Properties** information box

of positions for dimension figures are possible. Each of the **Measure** dimensioning tool icons has its own double square dot, showing that a *double-click* on any one of the icons will result in a dialogue box appearing on screen. No matter which of the five tool icons is selected, the **Dimension Settings** dialogue box appears (Fig. 8.6).

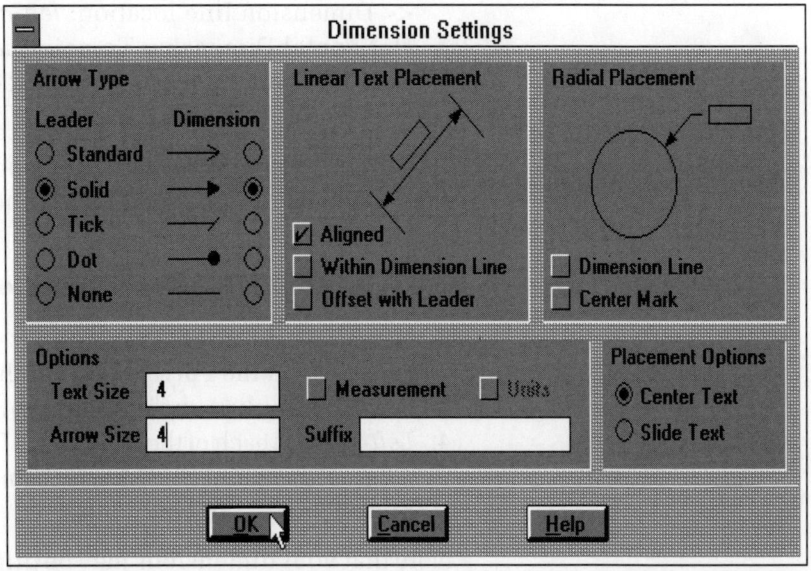

Fig. 8.6 The **Dimension Settings** dialogue box

Example 4 – the Dimension Settings dialogue box

1. *Double-click* on any one of the dimension tool icons in the **Measure** toolbox. The **Dimension Settings** dialogue box appears.
2. *Left-click* on each of the circles on each side of the **Solid** statement.
3. *Left-click* on the **Aligned** box.
4. *Enter* 5 in the **Text Size** box. The figure in the **Arrow Size** box adjusts automatically, although it can also be set independently.
5. Do not enter a suffix. If an abbreviation such as mm (millimetres), " (inches) or m (metres) is to be placed after every dimension, this suffix should be entered in the **Suffix** box.
6. *Left-click* on **OK**.

The dimensioning parameters are now set to solid arrows, with dimensions aligned with and above dimension lines, a text height of 5 (set to the text style selected in the **Text Style** dialogue box) and an arrow length of 4 mm.

When a dimension tool is in operation, prompts such as the following appear in the Prompt box:

Horizontal Dimension From point:
Horizontal Dimension To point:

When the **Offset With Leader** parameter is set (tick against the words in the **Dimension Settings** dialogue box), prompts such as the following appear after those when the **Offset With Leader** is not set:

Horizontal Dimension From point: *left-click* at required point
Horizontal Dimension To point: *left-click* at required point
<< Dimension line location: *left-click* at required point
Horizontal Dimension To point: *left-click* at required point
Text location: *left-click* at required point

The dimension figures appear at the selected point. A rubber-banded line from the centre of the dimension line appears on screen when the <<Dimension line location: point is being selected.

Example 5 – dimensions to the parameters already set

1. **New ...** the **A3SHEET.SKT**.
2. *Left-click* on the **Polyline** tool in the **Draw** toolbox.
3. Draw any outline such as that shown in Fig. 8.7.
4. *Left-click* on each of the **Horizontal Dimension**, **Vertical Dimension**, **Align Dimension** and **Angle Dimension** tools in turn and add dimensions as shown in Fig. 8.7.

Note that your dimensions may be different from those shown in Fig. 8.7 depending upon the outline you have chosen to draw.

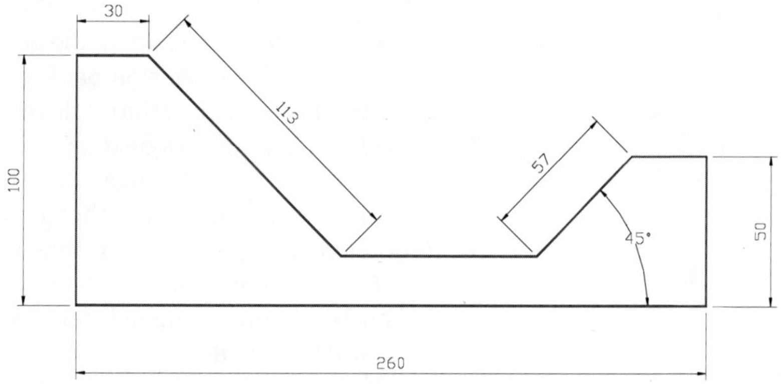

Fig. 8.7 Example 5

Example 6 – changing the dimension figure parameter

1. Draw any other outline, or repeatedly *left-click* on the **Undo** button until the dimensions on your drawing have been removed (undone).
2. *Double-click* on a **Dimension** tool to bring up the **Dimension Settings** dialogue box.

3. *Left-click* on the box next to **Within Dimension Line**.
4. Redimension your outline (Fig. 8.8).

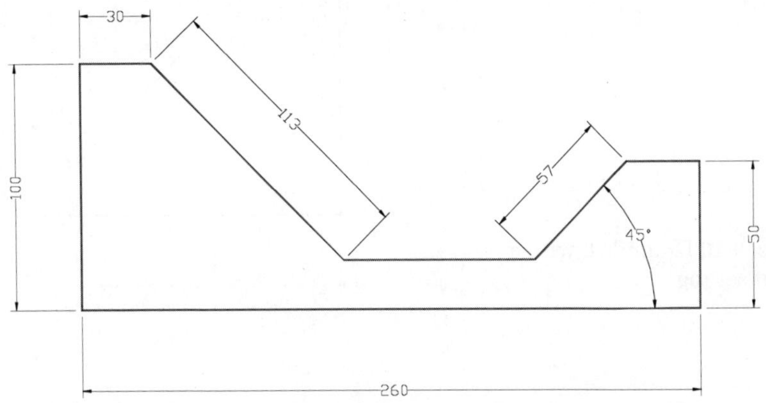

Fig. 8.8 Example 6

Example 7 – changing dimension arrows

1. Draw any other outline, or repeatedly *left-click* on the **Undo** button until the dimensions on your drawing have been removed (undone).
2. *Double-click* on a **Dimension** tool to bring up the **Dimension Settings** dialogue box.
3. *Left-click* on the box next to **Aligned**.
4. *Left-click* on both circles each side of **Tick**.
5. Redimension your outline (Fig. 8.9).

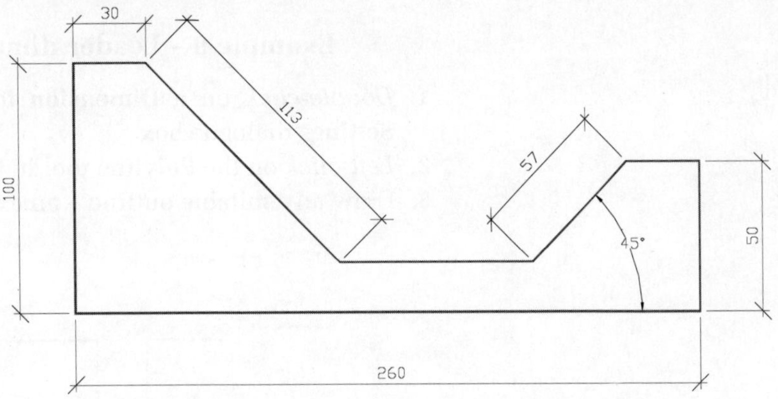

Fig. 8.9 Example 7

Example 8 – changing dimensioning parameters

Repeat the previous examples resetting **Dot** (Fig. 8.10) and resetting **Offset With Leader** (Fig. 8.11).

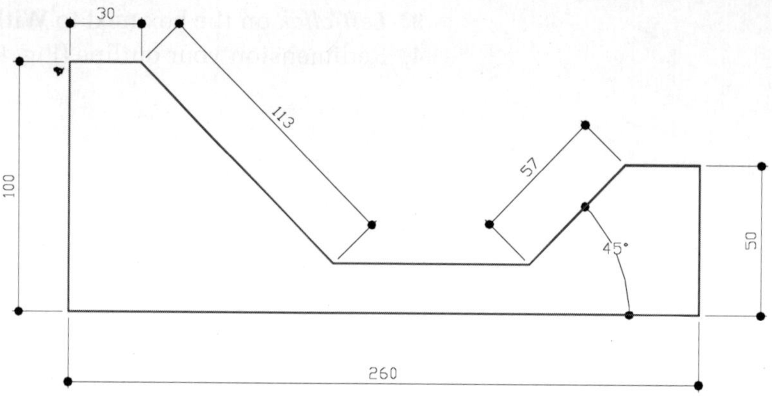

Fig. 8.10 Example 8 with a **Dot** setting

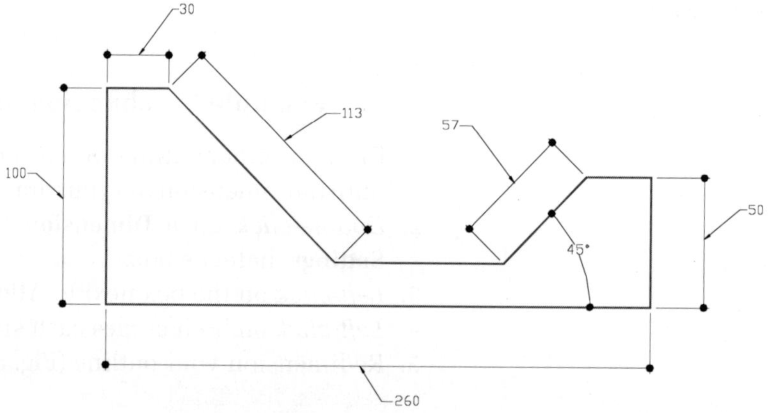

Fig. 8.11 Example 8 with an **Offset with Leader** setting

Example 9 – Leader dimensions

1. *Double-click* on a **Dimension** tool to bring up the **Dimension Settings** dialogue box.
2. *Left-click* on the **Polyline** tool in the **Draw** toolbox.
3. Draw any suitable outline – an example is given in Fig. 8.12.

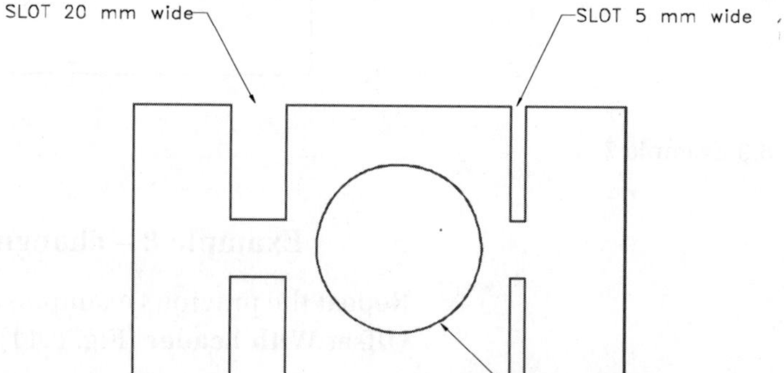

Fig. 8.12 Example 9

4. *Left-click* on the **Leader** tool in the **Measure** toolbox.

   Leader Start point: *left-click* at a suitable start point
   Leader To point: *left-click* at a suitable point
   Leader Text location *left-click* at a suitable point. The **Text Editor** dialogue box appears on screen

5. *Enter* the required text. An example is given in Fig. 8.13. *Left-click* on **OK**. The text appears at the selected text location point.

Note: The text entered in the **Text Editor** as shown in Fig. 8.13 includes %%c60. This results in the British Standard diameter symbol appearing before the figures as shown in Fig. 8.12. Several other %% calls are shown in Fig. 8.14.

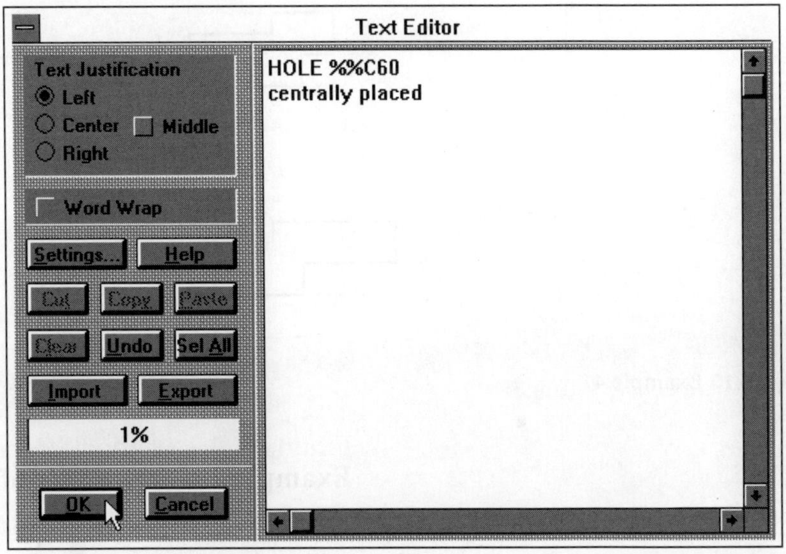

Fig. 8.13 The **Text Editor** dialogue box

%%O = OVERLINING (Start and finish with %%O)
%%U = UNDERLINING start and finish with %%U)
%%D = ° Degrees symbol
%%P = ± Tolerance symbol
%%C = ø Diameter symbol

Fig. 8.14 Special characters

Example 10 – the effect of Stretch on dimensions

1. **New ... A3SHEET.SKT**.
2. *Double-click* on **Polyline** in the **Draw** toolbox. *Enter* 0.7 in the **Polyline Width** box of the **Polyline Settings** dialogue box. Draw outline 1 of Fig. 8.15. Add its dimensions.
3. *Left-click* on **Multiple Copy** in the **Modify** toolbox.

Copy the outline several times on the screen.

4. *Left-click* on **Stretch** in the **Modify** toolbox. On the original outline you have drawn proceed as follows:

   Stretch First corner: *left-click* above the outline and between the right-hand end and the 30 dimension
   Stretch Crosses/window corner: *left-click* to the right and below the outline
   Stretch base: *left-click* on the bottom right corner of the outline
   Stretch to: *drag* the outline with its dimensions to the right

5. Repeat step 4 but selecting other points on the drawing to experiment with the results as in drawings 2, 3 and 4 of Fig. 8.15.

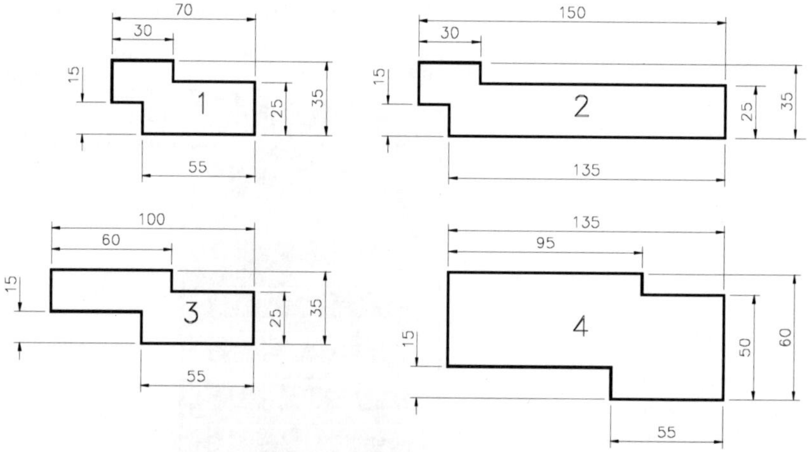

Fig. 8.15 Example 47

Example 11 – the effect of Scale on dimensions

Using the same outline as in Example 9, practise the effects of the tool **Scale** from the **Modify** toolbox on a drawing which includes dimensions.

Fig. 8.16 shows three examples of the expected results with scales of 0.6, 1.3 and 1.8.

Example 12 – the effect of Rotate on dimensions

Again, using the same outline, practise the effects of the **Rotate** tool from the **Modify** Toolbox.

Fig. 8.17 shows three examples of the expected results at various rotations. Note in particular drawing 4 showing a 90 degree rotation, in which the dimensions have become meaningless.

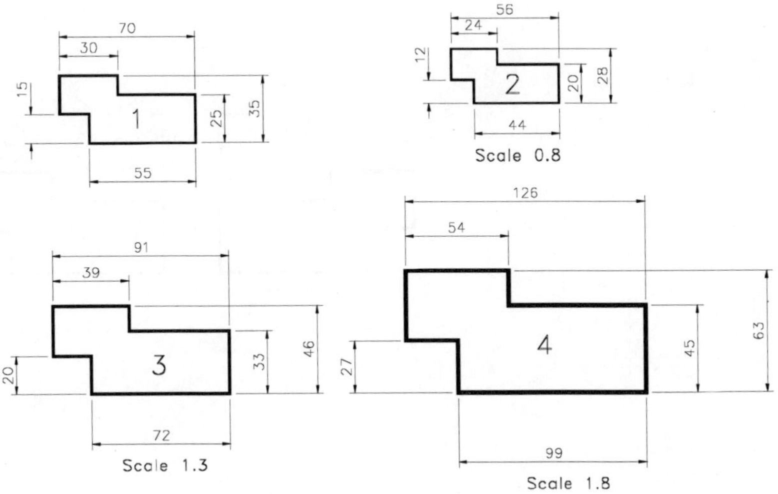

Fig. 8.16 Example 11

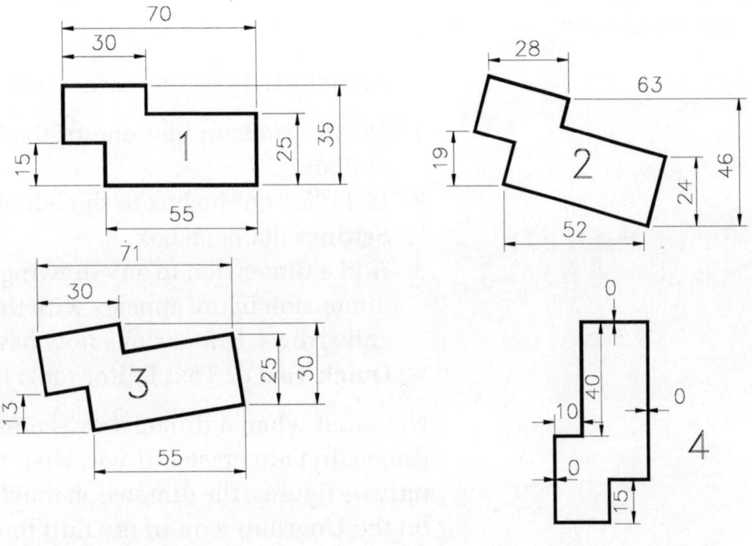

Fig. 8.17 Example 12

Example 13 – the effect of Mirror on dimensions

Again, using the same outline practice the effects of the **Mirror** tool from the **Modify** Toolbox.

Fig. 8.18 shows the results of using the tool both horizontally and vertically.

Associativity

From the above examples it can readily be seen why AutoSketch for Windows dimensioning is said to be **Associative**. The dimensions change according to the actions of tools from the **Modify** toolbox.

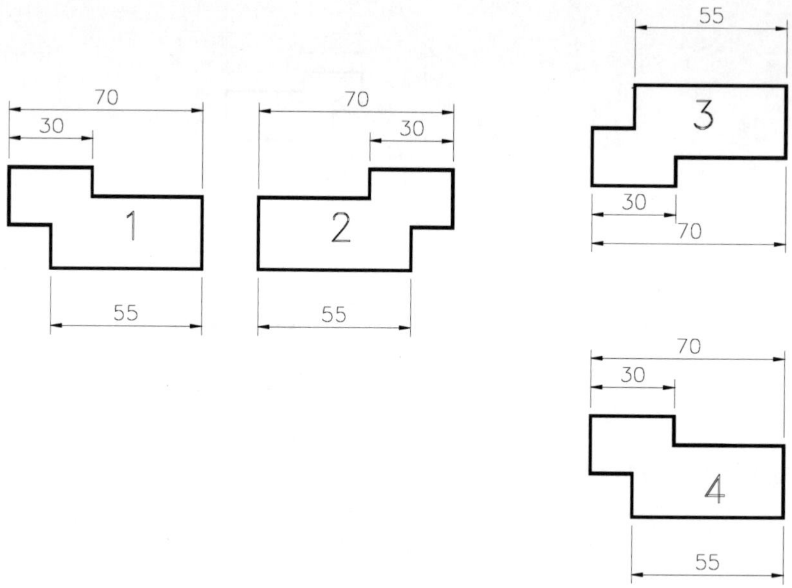

Fig. 8.18 Example 50

Associativity can be turned off by:

1. *Double-click* on any one of the **Dimension** tools in the **Measure** toolbox.
2. *Left-click* on the box to the left of **Measurement** in the **Dimension Settings** dialogue box.
3. Add a dimension to any drawing on screen. It will be seen that no dimension figure appears with the dimension line, extension lines and arrows. Dimensions now have to be added with the aid of the **Quick Text** or **Text Editor** tools from the **Draw** toolbox.

Note that when a dimension is erased, all of the objects within the dimension are erased. If you wish to erase part of a dimension, e.g. only its figures, the dimension must first be ungrouped by a *left-click* on the **Ungroup** icon in the **Edit** toolbox, followed by a *left-click* on the dimension. This explodes the dimension into its single objects, allowing any object within the dimension to be acted upon by any of the **Edit** tools.

In all the examples given here, the **Units** setting is at 0 decimal digits.

Exercises

1. Fig. 8.19. Turn off **Associativity** as described above. Then draw the given outline with **Polylines** of 0.7 width. Add the dimension lines, extension lines and arrows – but as **Associativity** is turned off, the figures will not appear. Add the dimension figures, with the aid of

the **Text Editor**. As an example the 130 dimension will be entered in the **Text Editor** as 130%%p0.05; the diameter of the circle will be entered as %%c30%%p0.05. Before the vertical dimensions can be added, the angle of the text must be altered in the **Text Settings** dialogue box – *enter* 270 against **Angle** in the **Text Options** area of the **Text Settings** dialogue box. Remember that the **Text Settings** dialogue box is called with a *double-click* on either the **Quick Text** or the **Text Editor** tool icons in the **Draw** toolbox.

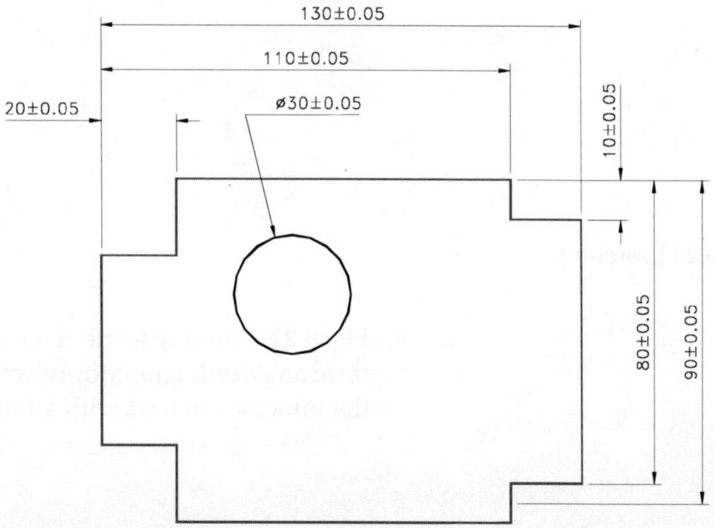

Fig. 8.19 Exercise 1

2. Fig. 8.20. Construct a third angle orthographic projection of the given component. Fully dimension your drawing.

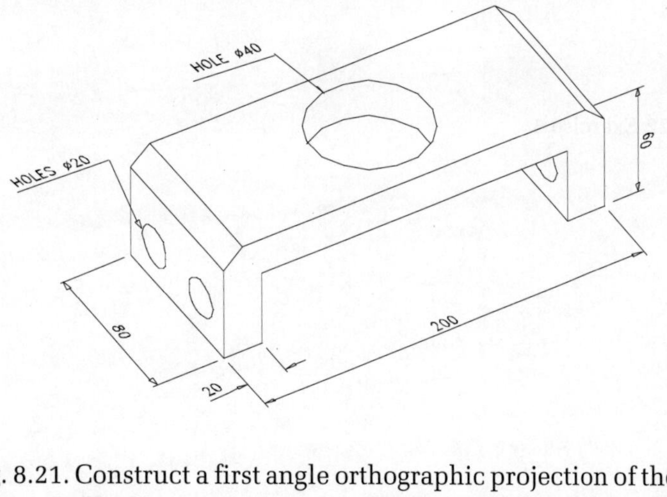

Fig. 8.20 Exercise 2

3. Fig. 8.21. Construct a first angle orthographic projection of the part shown. Fully dimension your drawing. Include the suffix mm with all dimensions.

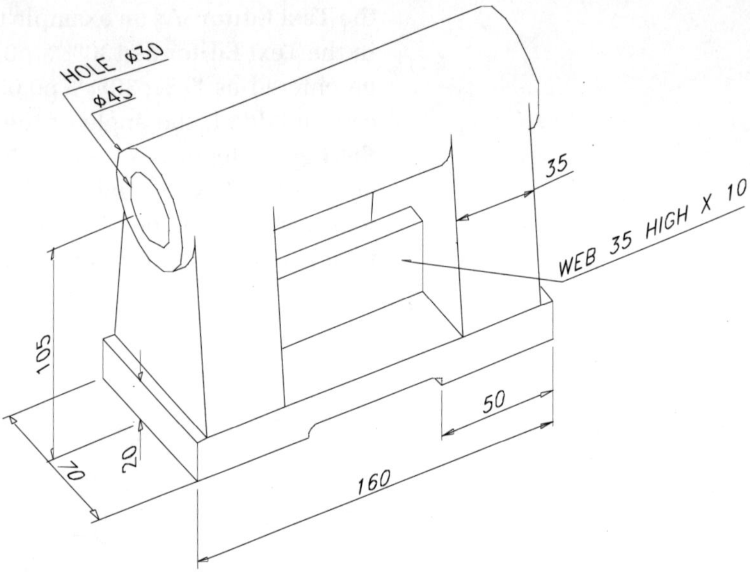

Fig. 8.21 Exercise 3

4. Fig. 8.22. Working to the given sizes, construct a fully dimensioned third angle orthographic projection of the given component. Include the tolerance of 0.05 with all dimensions.

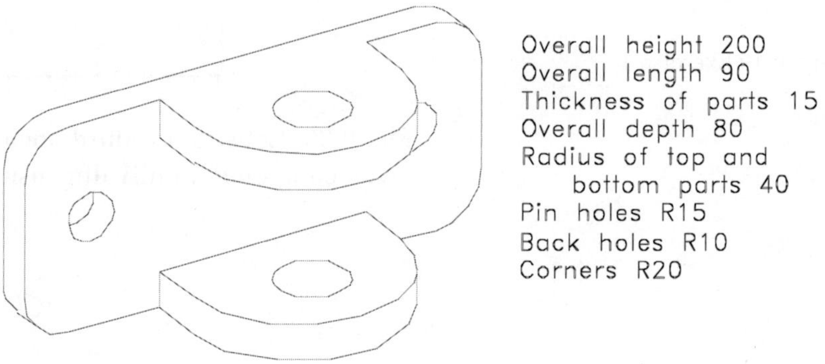

Fig.8.22 Exercise 4

CHAPTER 9

The Custom toolbox

Fig. 9.1 The **Custom** pull-down menu

Introduction

Fig. 9.1 shows the pull-down menu for the **Custom** toolbox and Fig. 9.2 for the icon toolbox. As can be seen from these two illustrations, a number of facilities are available with the aid of the tools in this toolbox, which are not mainly for the construction of drawings as such.

The Drawing Settings ... tool

Left-click on **Drawing Settings ...** in the **Custom** toolbox. The **Drawing Settings** dialogue box appears on screen. As can be seen in Fig. 9.3, any one of the various drawing parameters can be set from this dialogue box. Fig. 9.3 shows the results of a *left-click* on **Attach ...** in the dialogue box – bringing a second dialogue box on screen – for **Attachment Settings.** As shown, **Attach Mode** has been set on (tick in box against **Attach Mode**) and only **End Point** and **Intersect** have been made active.

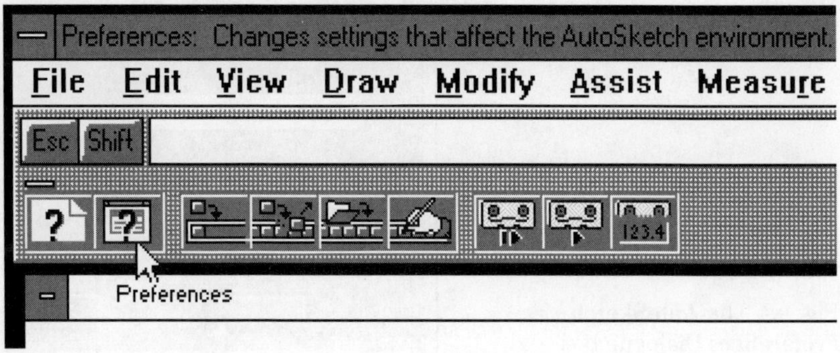

Fig. 9.2 The **Custom** icons toolkit

The Preferences ... tool

Left-click on **Preferences ...** and the **AutoSketch Preferences** dialogue box appears. As seen from Fig. 9.4, **Undo** and **Selection Highlighting**

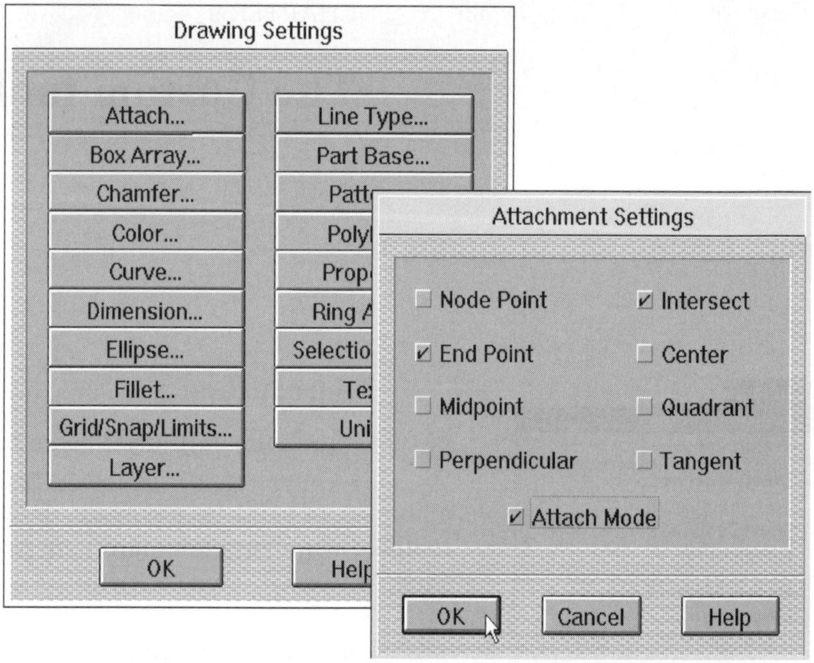

Fig. 9.3 The **Drawing Settings** dialogue box

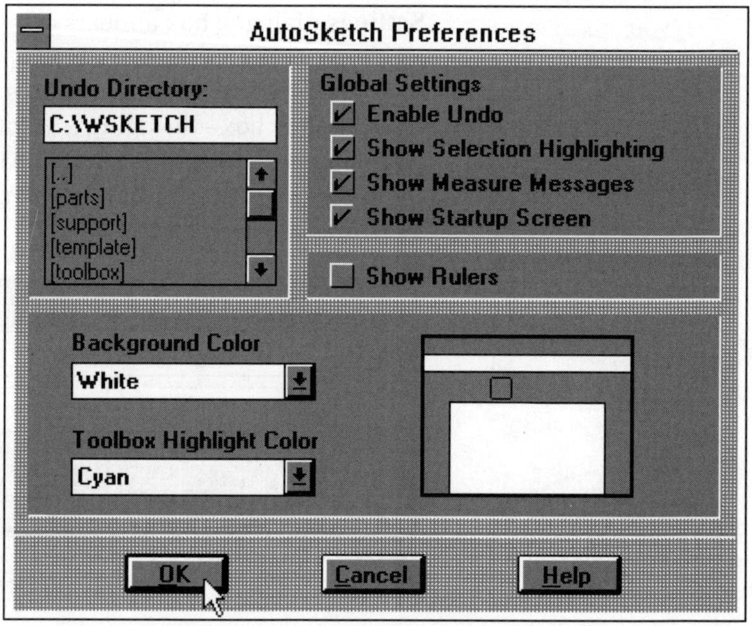

Fig. 9.4 The **AutoSketch Preferences** Dialogue box

can be disabled and the **Background** screen colour can be changed from within this dialogue box.

Example 1 – the Toolbox Editor

1. *Left-click* on **Create Toolbox** in the **Custom** toolbox. The **Toolbox Editor** appears on screen – Fig. 9.5.
2. Following the instruction in the dialogue box **To Add a Button**, *drag* tool icons from various toolboxes to make up your own toolbox. As can be seen from Fig. 9.5, those chosen have been the tools most frequently used for constructing a drawing. However you can make up your own toolbox as you wish from existing tool icons.
3. *Left-click* on **Save as ...** and *enter* a toolbox name in the dialogue box which appears. I have entered **AY** as the name of my toolbox. Its full file name will be **ay.mcr**.
4. Experiment with the options available for rearranging or deleting buttons while in the **Toolbox Editor**.

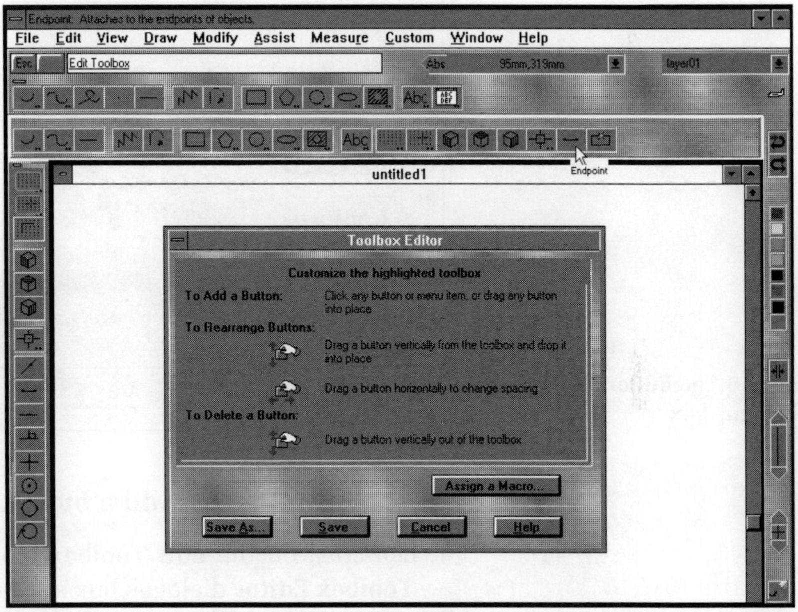

Fig. 9.5 The **Toolbox Editor** dialogue box

Example 2 – the Button Editor

1. *Left-click* on the **Button Editor** icon in the **Utilities** toolbox. The **Button Editor** dialogue box appears on screen – Fig. 9.6.
2. *Left-click* on each of the icons in buttons at the top of the editor. The drawing tool within the editing box in the centre of the dialogue box changes according to whether a pixel tool (pencil shape icon), a line drawing tool, a circle/ellipse drawing tool or a rubber are selected.

3. Experiment inside the central editing box with each drawing tool in turn. Fig. 9.6 shows my initials A.Y worked in the box by placing the pencil-like icon on pixels, then giving a *left-click*. If the pixel is grey, it turns black; if black, it turns grey. Patterns, shapes, icons can be drawn in pixels in this way. Note that a small box to the scale that the button will appear on screen shows the actual pixel drawing as the work proceeds.
4. *Left-click* on **Save As ...** and save your button to a file name. In this example it has been saved to the file name **AY.BMP**.

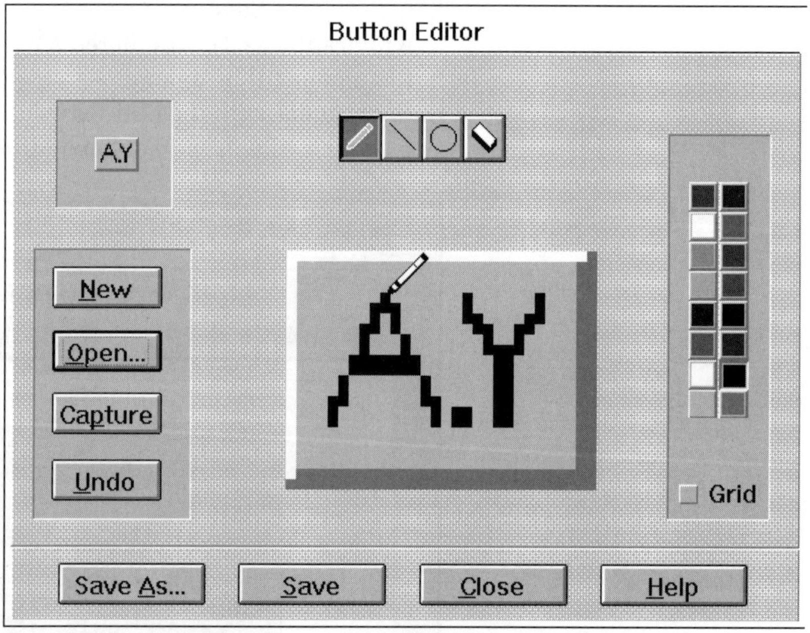

Fig. 9.6 The **Button Editor** dialogue box

Example 3 – Add a button to a toolbox

1. *Left-click* on the **Edit Toolbox** icon in the **Custom** toolbox. The **Toolbox Editor** dialogue box appears.
2. *Left-click* on **Assign a Macro** in the dialogue box. The **Assign a Macro** dialogue box comes onto screen – Fig. 9.7.
3. *Left-click* on **ay.mcr** and on **ay.bmp**. These filenames are then repeated in the **Macro** and **Button** areas of their dialogue box.
4. *Left-click* on **Add** and the **AY** button joins those in the **AY** toolbox – Fig. 9.8.
5. The toolbox, now with the **AY** added, is saved as toolbox **AY** in the **Toolbox Editor** – Fig. 9.8.

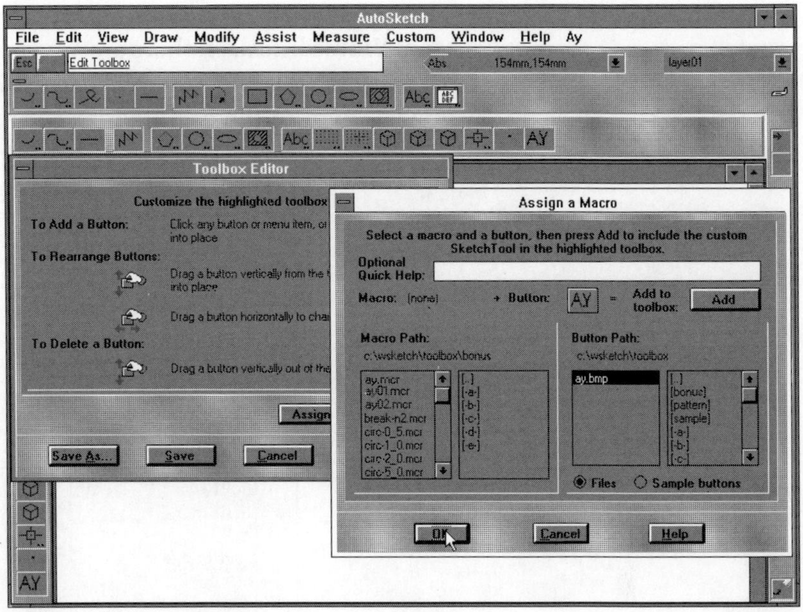

Fig. 9.7 The **Toolbox Editor** together with the **Assign a Macro** dialogue boxes

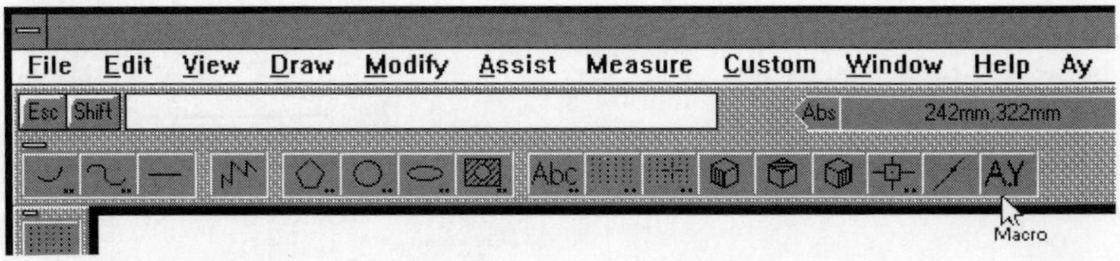

Fig. 9.8 The new toolbox **AY**

Example 4 – Record and Play a Macro

1. *Left-click* on the **Record Macro** icon in the **Custom** toolbox. The **Record Macro** dialogue box appears on screen – Fig. 9.9.
2. *Enter* a file name in the **Filename:** box – in the example given this is **drawings**.
3. **Open ...** a number of already drawn AutoSketch drawings, one after the other.
4. *Left-click* on the **End Macro** icon in the **Custom** toolbox (this icon had previously been the **Record Macro** – it changed to **End Macro** when it was first selected).
5. *Left-click* on **Play Macro** icon in the **Custom** toolbox. *Left-click* on **drawings.mcr** in the file list.
6. The drawings which were opened one after the other are repeated by appearing on screen one after the other in the same order as they were recorded.

Note that with a *left-click* on **File** in the **Menu Bar**, you will see the tool **View Slide ...** . If there are any slide files on your disks, slides can also be included in **Macros**. So can **.dxf** files opened with the **Import DXF ...** tool in the **File** toolbox.

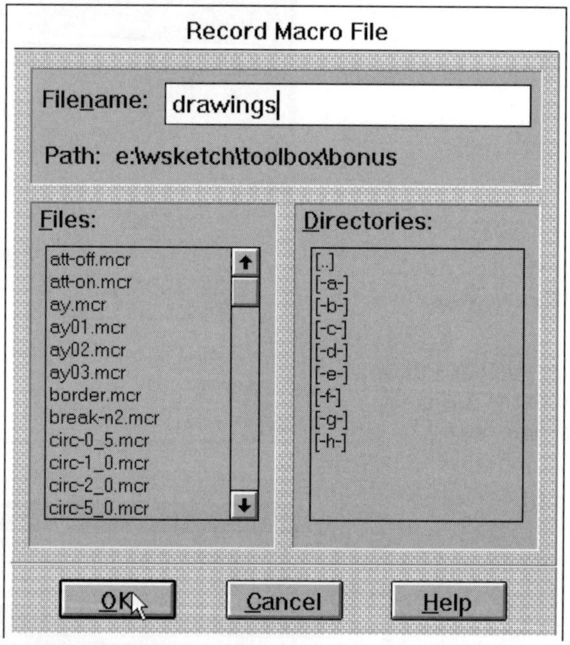

Fig. 9.9 The **Record Macro File** dialogue box

Example 5 – Copy to Windows Bitmap files

1. **Open ...** any drawing file – in this example it is plans of the two storeys of a house, Fig. 9.10.
2. *Left-click* on the **Copy as Bitmap** icon in the **Edit** toolbox:

   Copy Bitmap First corner: *left-click* at an appropriate position
   Copy Bitmap Second corner: *left-click* at an appropriate position

3. Either *left-click* on the **Show Clipboard** icon in the **Edit** toolbox, or *left-click* in the Windows button in the top left of the window. This brings a pull-down menu on screen. *Left-click* on **Clipboard** in the pull-down menu. The bitmap appears in the Clipboard Viewer (Fig. 9.11).

Example 6 – Paste objects from a drawing

1. *Left-click* on the **Copy Objects** icon in the **Edit** toolbox.

   Copy Objects Select object: *left-click* to the left and above the area of the drawing to be copied

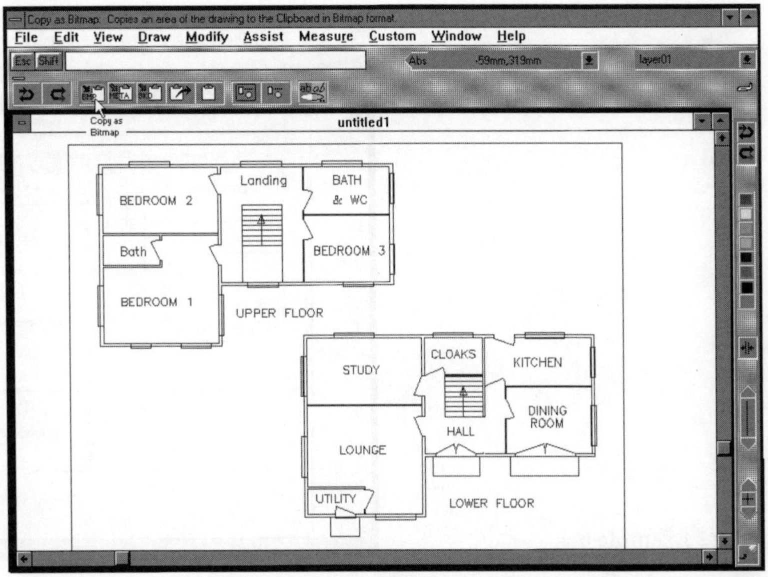

Fig. 9.10 Example 5 – the drawing chosen for making a bitmap

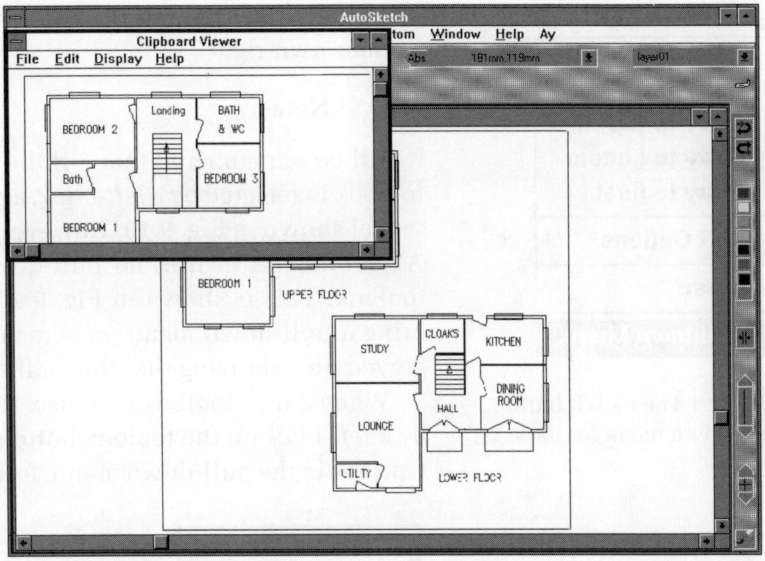

Fig. 9.11 Example 5 – The **Clipboard Viewer**

Copy Objects Crosses/window corner: *left-click* to the right and below the area of the drawing to be copied

2. *Left-click* on the **Paste** icon in the **Edit** toolbox.
3. The previously copied area of the drawing appears on screen attached to the mouse cursor.
4. *Drag* the objects to the required position on screen and *left-click*. The objects are pasted on screen – Fig. 9.12.

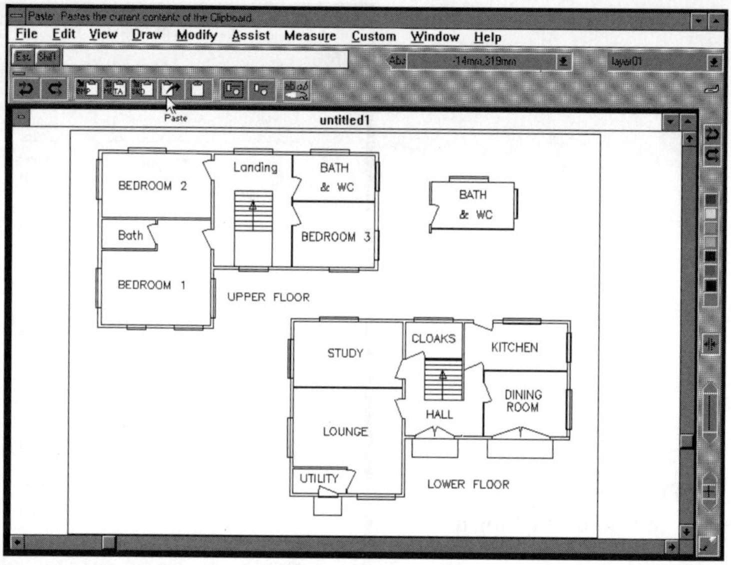

Fig. 9.12 Example 6

Note: Objects copied from a drawing in this manner can be pasted into other drawings if needed, or can be pasted on screen as drawings in their own right.

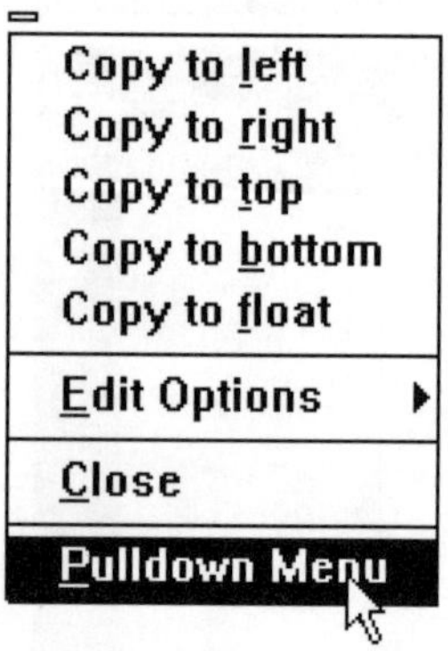

Fig. 9.13 The toolkit button pull-down menu for the **AY** toolkit

Notes

It will be remembered that – if the button at the top left corner of a toolbox is selected by a *left-click*, a pull-down menu appears, similar to that shown in Fig. 9.13. However, when a new toolbox such as my **AY** toolbox is in use, no pull-down menu is associated with the toolbox. This is shown in Fig. 9.14 – an attempt has been made to bring a pull-down menu on screen, but the words in the menu are greyed out, showing that the facility is not available.

When a new toolbox is in use, it can be removed from the screen by a *left-click* on the toolbox button, followed by a *left-click* on **Edit Options** in the pull-down menu, followed by yet another *left-click* on

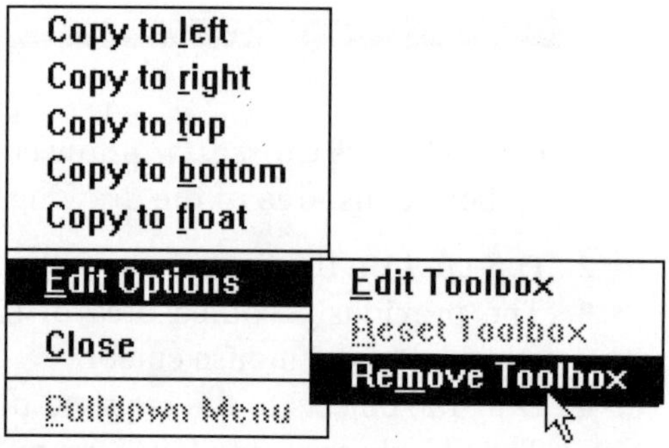

Fig. 9.14 The **Remove Toolkit** function from the toolkit button pull-down menu

Remove Toolbox in the sub-menu which appears. These menus are shown in Fig. 9.14.

Other forms of macro are not dealt with in this book. AutoSketch for Windows will allow the compilation of ASCII files in a text editor, which will automate some actions. A number of these ASCII files will be found in the directory **wsketch\toolbox\bonus**. The reader might wish to examine these files in the MS-DOS text editor **edit**. It will be seen that macro files have their own language which, while not difficult to use, is beyond the scope of a book of this nature.

Some details about **Slide** files and **DXF** files will be given in a later Chapter 12.

CHAPTER 10

Project suggestions

Introduction

This chapter contains a series of ideas for projects, all of which incorporate some element of design. The projects vary from comparatively simple to more complex. The reader is advised to attempt the design parts of the projects. If these are found to be too difficult or time-consuming, then at least use the skills in drawing with AutoSketch for Windows which you have acquired so far by merely copying the given drawings.

Project 1

Fig. 10.1 is a design of a logo for a magazine publishing firm known as Pentagon Limited. First make a copy of the given logo. Then:

(a) Design and construct a second logo based on a pentagon.
(b) Design a logo based on a hexagon.

Work to any convenient sizes. The external angle of a regular pentagon is 72 degrees and of a regular hexagon is 60 degrees.

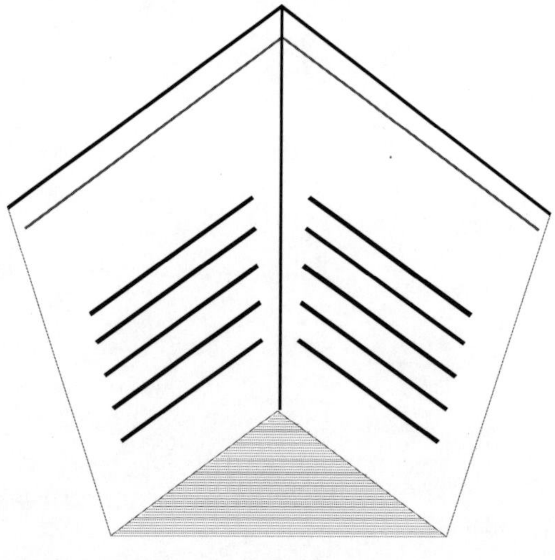

Fig. 10.1 Project 1

Project 2

A firm, Bike Limited, deals in motor cycles and shows the logo given in Fig. 10.2 on the sides of the tanks of the motor cycles it sells.
(a) Copy the given logo.
(b) Design and construct another logo which the firm could use.

Project 3

Fig. 10.3, 10.4 and 10.5 give details of the layout of the fittings and furniture in a kitchen. The elevations of the walls have been drawn as if they had been laid flat outwards from their normal positions.

The given layout could be improved in a number of ways. Working on an AutoSketch for Windows screen set up for an A2 sheet (**Limits** set to 0,0 and 594 × 420), re-draw the layout so as to achieve a better designed position for the fittings and furniture in the kitchen.

Project 4

Fig. 10.6 is an example of a pattern constructed with the aid of the tools **Polyline** (set to 10 wide and with a Pattern Fill), **Arc**, **Circle** and **Ring Array**, working in AutoSketch for Windows.

Fig. 10.7 is an example of a pattern constructed with the aid of the tools – **Curve**, **Mirror**, **Multiple Copy**, **Ellipse** and **Arc**, working in AutoSketch for Windows.

Fig. 10.8 is an example of a pattern constructed with the aid of the tools – **Circle**, **Curve**, **Ring Array**, **Polyline** (set to 5 wide and with a Pattern Fill), and **Break**, working in AutoSketch for Windows.

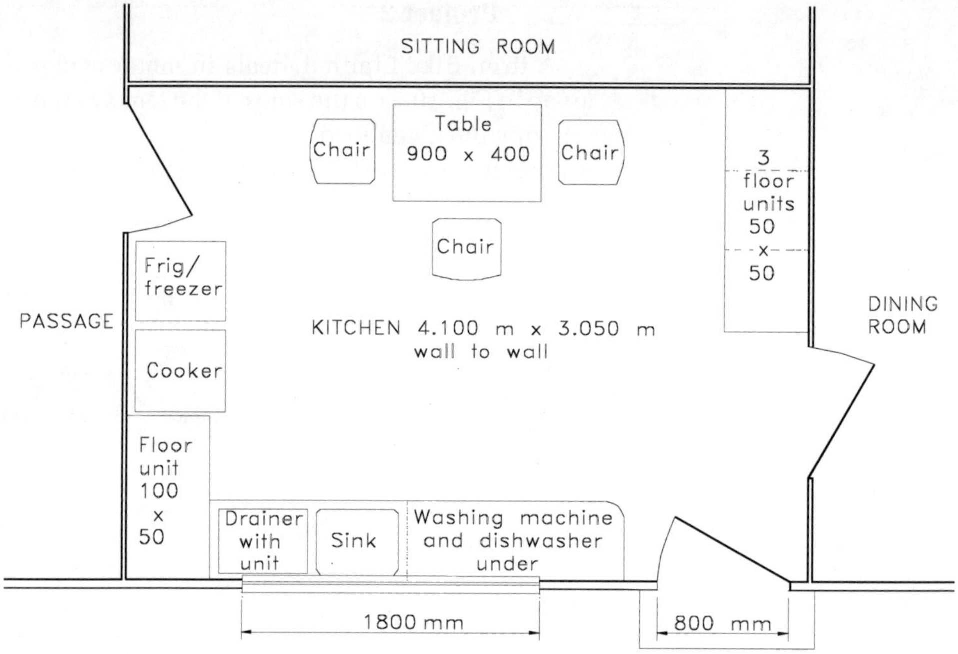

Fig. 10.3 Project 3 – plan of kitchen

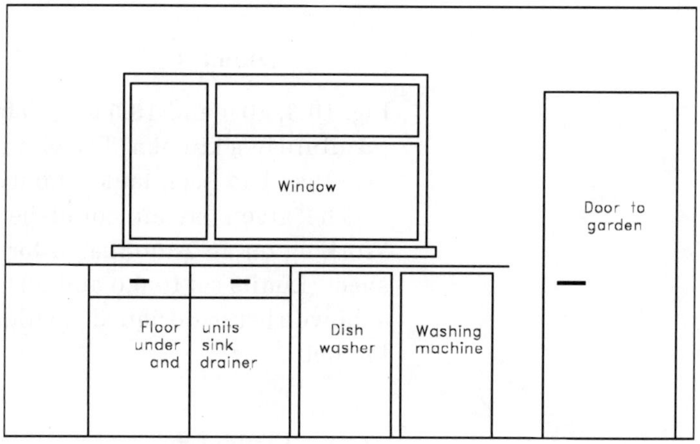

Fig. 10.4 Project 3 – outer wall of kitchen

Design a pattern of a similar nature. When you have designed the pattern, repeat it with the aid of **Box Array**, **Ring Array** or **Multiple Copy**.

Project 5

Fig. 10.9 is a plan of a rectangular dining table set for four people. Design and construct a plan drawing in AutoSketch for Windows of a setting for six people around an elliptical table. The size of the

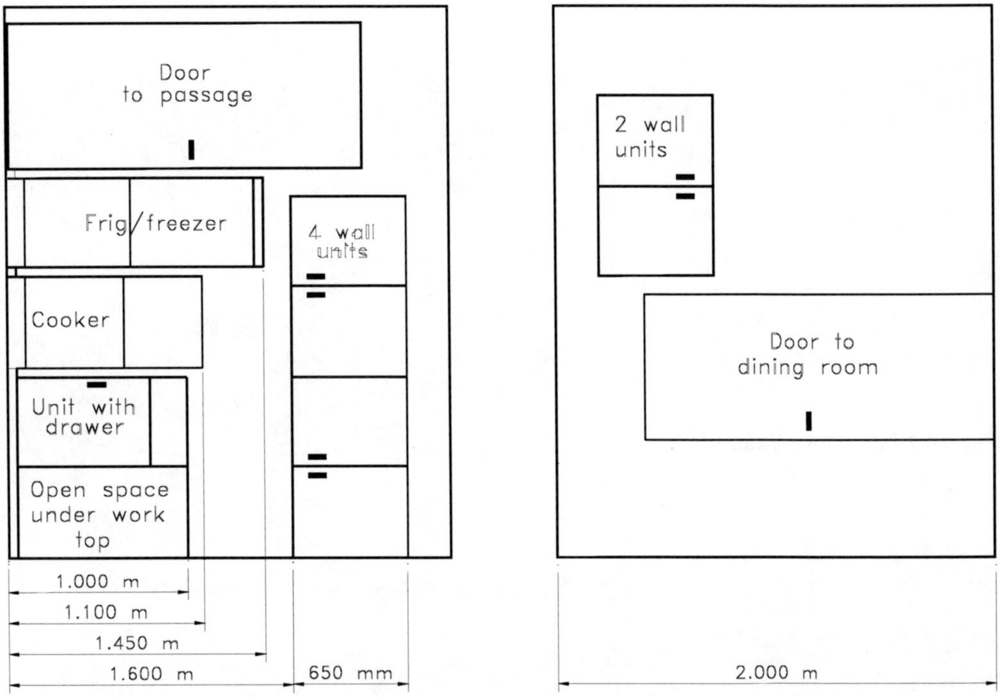

Fig. 10.5 Project 3 – walls of kitchen

Fig. 10.6 Project 4 – Example 1

Fig. 10.7 Project 4 – Example 2

Fig. 10.8 Project 4 – Example 3

table will depend upon the sizes of crockery and cutlery. Fig. 10.10 shows the table setting used in Fig. 10.9.

Fig. 10.9 Project 5

Fig. 10.10 Project 5 – suggestion for a table setting

Project 6

Fig. 10.11 is a first angle orthographic projection of a tiled table. Construct the given drawing within **Limits** 0,0 and 594,420 (A2 sheet sizes) but with a different design for the tiles set in the table top.

Drawings saved in AutoSketch for Windows

Fig. 10.12 shows the drawings saved in AutoSketch for Windows as shown in the **Open Drawing File** dialogue box. Note that **fig05.skd** is missing because it had not been drawn at the time this dialogue box was opened.

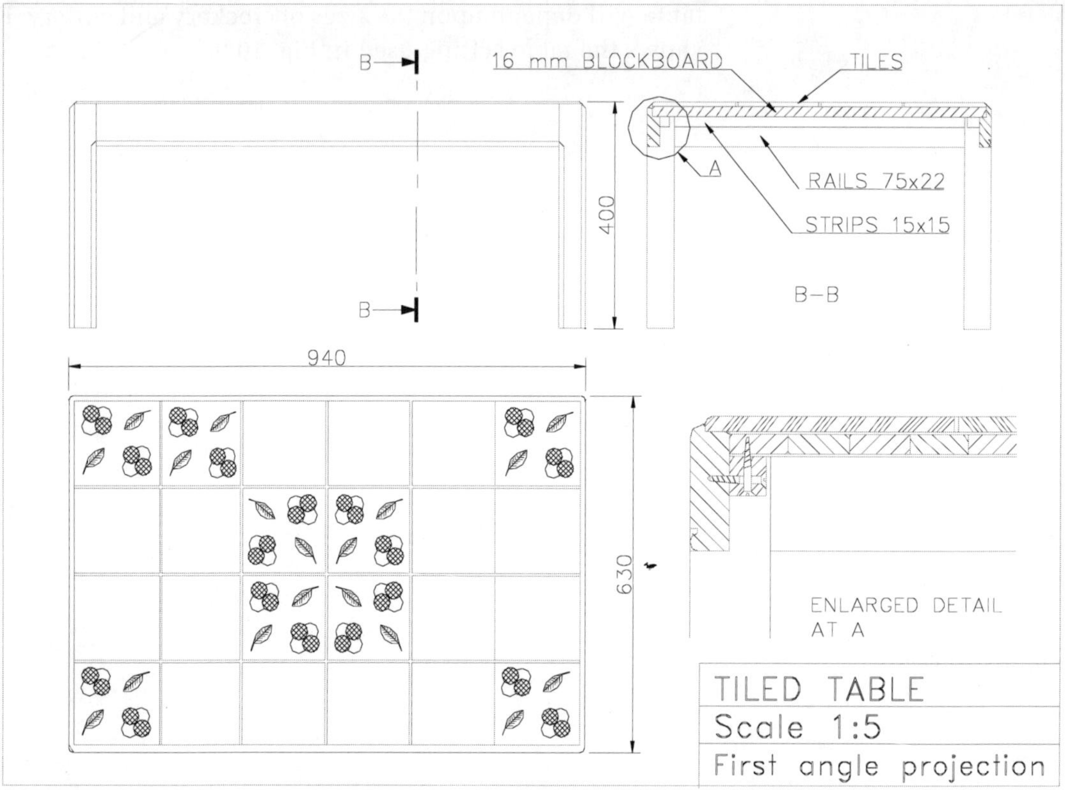

Fig. 10.11 Project 6

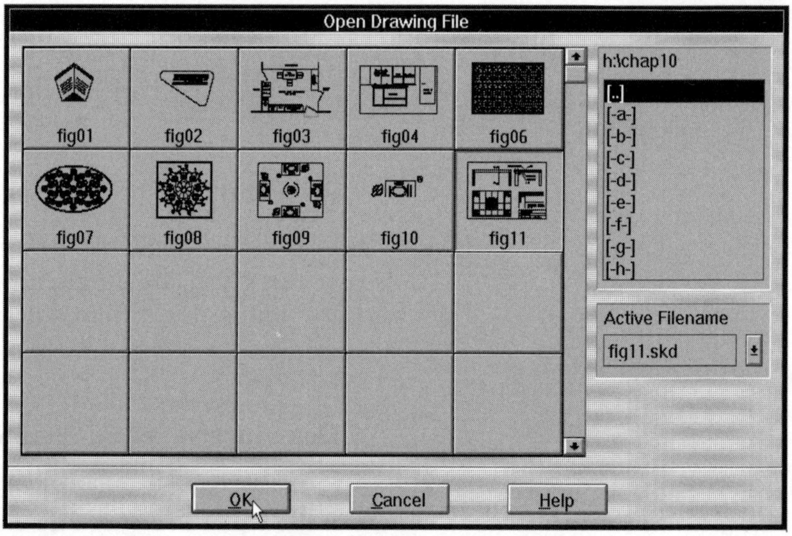

Fig. 10.12 The **Open Drawing File** dialogue box

CHAPTER 11

MS-DOS and Windows

Introduction

In this book we are primarily concerned with AutoSketch for Windows, i.e. with the AutoSketch software installed in a 386 or 486 computer running under MS-DOS and Windows. In this chapter some information regarding commands in MS-DOS and Windows as they affect the use of AutoSketch for Windows is briefly described. Although AutoSketch for Windows will run with any MS-DOS system from version 3.3 onwards, its best performance will be with MS-DOS version 5.0 or higher. The software will run in Windows 3.0, but better results will be achieved if Windows 3.1 or higher is used. Both MS-DOS and Windows are complex software programs and this chapter gives only a very brief summary of a few of the numerous commands available with the programs.

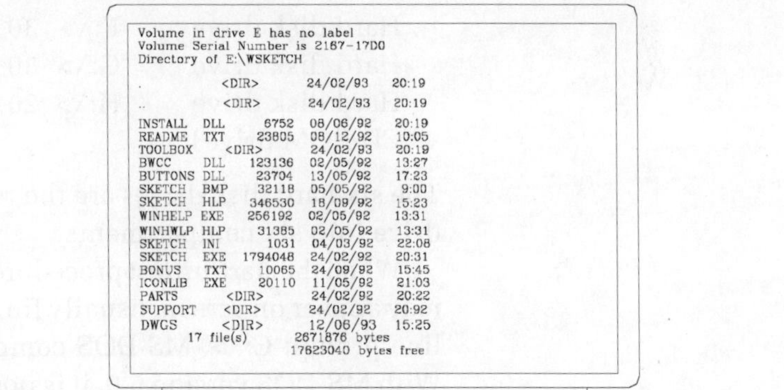

Fig. 11.1 A typical PC setup such as would be used with AutoSketch for Windows, showing the directory for **WSKETCH**

MS-DOS is an abbreviation for Microsoft Disk Operating System. Microsoft Corporation is an American software company. Both Windows and MS-DOS are software products of the Microsoft Corporation.

Start-up

When a PC (Personal Computer) running under MS-DOS is switched on, a self-test is run – memory, system circuits, disk drives and other peripherals are checked. When satisfied that these are OK, some of the MS-DOS files are loaded into memory. At this stage only those commands most frequently used are loaded. Commands not used so often are loaded as and when they are required.

The PC may have several drives – see Fig. 11.1. AutoSketch for Windows files must be loaded on a hard disk, so in the example given in Fig. 11.1, the files will be on drive E:\>. PCs may have more drives than those shown in Fig. 11.1 – up to as many as 26 are possible working under MS-DOS, although it is more usual to have five or six, e.g. floppy disk drives A:\ and B:\, hard disk drives C;\, D:\ and E:\, say, and perhaps a CD-ROM drive F:\. The computer on which this book was written has disk drives as follows:

Floppy disk drive	A:\>	$3\frac{1}{2}$ inch 1.44 Mbyte
Floppy disk drive	B:\>	$5\frac{1}{4}$ inch 1.2 Mbyte
Hard disk drives	C:\>	60 Mbytes
Hard disk drive	D:\>	30 Mbytes
Hard disk drive	E:\>	30 Mbytes
Hard disk drive	F:\>	30 Mbytes
Hard disk drive	G:\>	30 Mbytes
Hard disk drive	H:\>	20 Mbytes
CD-ROM drive I:\>		

The six hard disk drives are the result of dividing a 210 Mbyte hard drive into six compartments.

When the start-up test procedure is completed, a series of statements may appear on screen, usually finishing with the MS-DOS command line prompt **C:\>**. MS-DOS commands are entered at this prompt. With MS-DOS version 5.0, it is possible to have the start-up showing a graphics screen from which commands, etc., can be selected with the aid of a mouse. Here we will only be concerned with commands entered from the keyboard.

When a PC is switched on from the power switch, this is said to be a *cold start*. When the computer has been running for some time, if a fault occurs a *warm start* can be made by pressing the three keys **Ctrl/Alt/Del**. It is advisable to avoid too many cold starts when using a PC, because repeated cold starts may damage the chips in the system circuits. Warm starts do not carry this risk. Avoid warm starts while working in Windows if possible – it is possible to lose some of your construction if you do so. It is a good practice in any case to save your work at regular intervals – say every 15 minutes or so. Then if anything goes wrong, e.g. a fault in the power supply, at least most of your work will have been saved.

Some MS-DOS commands

Either capital or lower-case letters can be used for any of the MS-DOS commands.

To change from drive C:\> to drive E:\>

C:\> e: *Return*
E:\>

To see the sub-directories and files in a drive

To see the contents of the directory **WSKETCH** in which the files for AutoSketch for Windows are held, as shown in Fig. 11.2:

E:\> cd wsketch *Return*
E:\WSKETCH> dir *Return*

Taking examples from the above directory:

Volume name is that given to the hard disk by the operator. In the example given, no **Volume** name was entered
Volume serial number is automatically given to a disk by MS-DOS version 5.0 (or later)
WSKETCH is the directory holding the AutoSketch files
<DIR> shows that this is the name of a directory
19/08/92 shows that the directory was made on 19 August 1992
11.25 is the time of day (24-hour clock) when the directory or file was saved to disk
ICONLIB is the name of a file which holds the library of icons for AutoSketch for Windows, **EXE** being the file extension name
20110 against the file name and its extension is the number of bytes in that file

To list files in a directory

To give a list of all sub-directories and files in the directory **e:\wsketch** as shown in Fig. 11.2:

E:\> dir wsketch *Return*

```
Volume in drive E has no label
Volume Serial Number is 2167-17D0
Directory of E:\WSKETCH
.              <DIR>        24/02/93    20:19
..             <DIR>        24/02/93    20:19
INSTALL  DLL      6752      08/06/92    20:19
README   TXT     23805      08/12/92    10:05
TOOLBOX        <DIR>        24/02/93    20:19
BWCC     DLL    123136      02/05/92    13:27
BUTTONS  DLL     23704      13/05/92    17:23
SKETCH   BMP     32118      05/05/92     9:00
SKETCH   HLP    346530      18/09/92    15:23
WINHELP  EXE    256192      02/05/92    13:31
WINHWLP  HLP     31385      02/05/92    13:31
SKETCH   INI      1031      04/03/92    22:08
SKETCH   EXE   1794048      24/02/92    20:31
BONUS    TXT     10065      24/09/92    15:45
ICONLIB  EXE     20110      11/05/92    21:03
PARTS          <DIR>        24/02/92    20:22
SUPPORT        <DIR>        24/02/92    20:92

         17 file(s)      2671876 bytes
                        17623040 bytes free
```

Fig.11.2 The MS-DOS directory and files structure

Directories, sub-directories and files

Fig. 11.3 shows the structure of the MS-DOS disk filing hierarchy. In the examples given in Fig. 11.3:

E:\> is the name of the drive holding the directories and files
WSKETCH is the name of the directory holding the AutoSketch for Windows system files
TOOLBOX is the sub-directory holding the AutoSketch for Windows toolbox files
SUPPORT is the sub-directory holding AutoSketch for Windows support files
PARTS is the sub-directory holding the parts files
DWGS is a sub-directory of **WSKETCH**, which itself has three sub-directories:

AY is the sub-directory of **DWGS** in which my drawing files (those constructed by A. Yarwood) are held
RBO is the sub-directory of **DWGS** in which the drawing files of the operator whose initials are RBO are held
BHL is the sub-directory of **DWGS** in which the drawing files of the operator whose initials are BHL are held

To make a directory

E:\> md (or mkdir) wsketch *Return*
E:\>

and the directory **WSKETCH** is formed on drive **E:\>**.

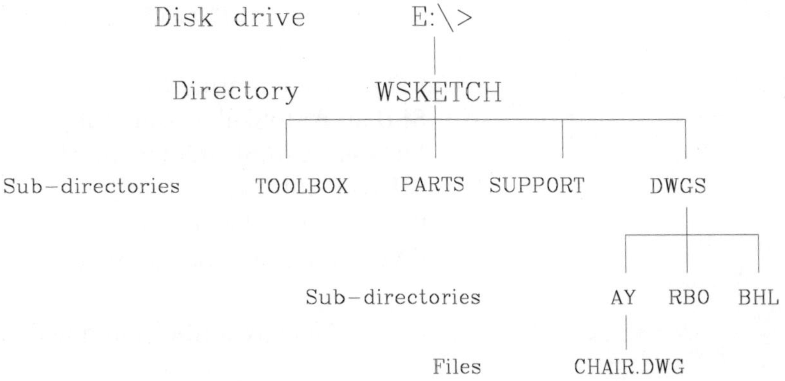

Fig. 11.3 The sub-directories and some files in the directory **WSKETCH**

To change directories

E:\> cd wsketch *Return*
E:\WSKETCH>

To make a sub-directory in WSKETCH

E:\ WSKETCH> md dwgs *Return*
E:\>

File names and extensions

File names usually have a file extension. The file **chair** in sub-directory **ay** will have a file name extension **.skd**. Thus its full name is:

E:\WSKETCH\DWGS\AY\CHAIR.SKD

Note the back-slashes between directory and sub-directories (or file names) and the fullstop between the file name and its extension. These are obligatory.

Extension examples

A number of file name extensions are used in AutoSketch for Windows files. Those in most common use are:

.SKD extension for an AutoSketch drawing file
.SKT extension for AutoSketch template file
.BAK extension for a backup file. If an AutoSketch drawing file is saved a second (or more) times a backup file (extension **.bak**) will form automatically on disk
.EXE (or **.COM**) files which execute a software program. An example is **sketch.exe** held in the directory **e:\wsketch**. When AutoSketch

is called, **SKETCH.EXE** loads into memory and the program commences

.CFG a configuration file
.SLD an AutoSketch slide file
.MCR an AutoSketch macro file
.SHX an AutoSketch font file
.PAT an AutoSketch pattern file
.DXF a data interchange file (or Data eXchange File)

To copy a file from one directory to another

E: copy wsketch\dwgs\ay\chair.skd wsketch\dwgs\draw\rbo *Return*

To copy a number of files between directories

To copy all files in the sub-directory **AY** to **RBO**:

E:\\> copy\wsketch\ay\draw*.* wsketch\dwgs\rbo *Return*

To rename a file

To rename the file **CHAIR.SKD** in the sub-directory **AY** as **ARMCHAIR.SKD**

E:\\>renamewsketch\dwgs\ay\chair.skd wsketch\dwgs\ay\armchair.skd *Return*

To erase a file

E:\\> erase (or del) wsketch\dwgs\ay\chair.skd *Return*

Note: Take great care when erasing. If a backup file (extension **.bak**) has replaced the erased file, it will be possible to rename it to the file name it is backing up. Otherwise a file may be lost. MS-DOS version 5.0 does however have an **UNDELETE** command, which may save such a situation.

To send the contents of a file to a printer

E:\\> type wsketch\readme.txt > prn *Return*

or

E:\\> wsketch\readme.doc > prn *Return*

Note that this form of usage is not of any use for most files from software packages because they will probably be written in machine code. Only files such as those ending in **.txt** or **.doc**, other text files or those written in ASCII code are suitable for this usage.

Windows

Usually Windows is opened by *entering* **win** at the **C:\>** MS-DOS prompt. However, it may be necessary to change to the Windows directory before *entering* the **win**. Normal practice is to have a *batch* file (extension **.BAT**) in the **C:\>** drive to start up Windows.

If the batch file does not exist, *enter* the following to create it:

C:\> copy con win.bat *Return*
cd windows *Return*
win *Return*
cd\ *Return*
Ctrl\Z *Return*

This changes to the Windows directory and starts up Windows. The **cd** at the end of the batch file will change the system back to **C:\>** when coming out of Windows.

When Windows appears, the screen will look something like that shown by Fig. 11.4. The appearance of the Windows start-up screen depends upon the software loaded into Windows. Note that one of the icons on screen is that for AutoSketch. A *double-click* on the icon brings AutoSketch for Windows on screen, preceded by the coloured icon shown by Fig. 11.5. This icon stays on screen for a few seconds, at the end of which the full AutoSketch screen appears.

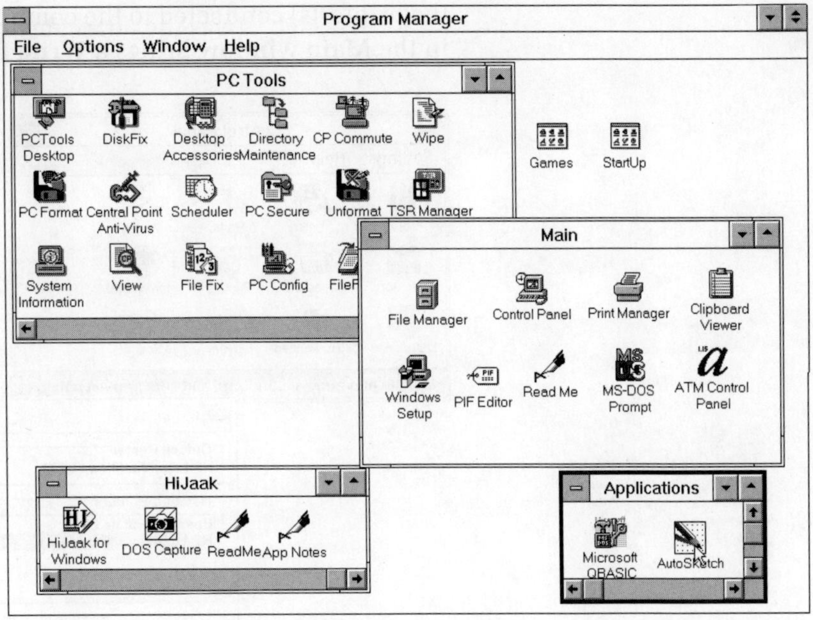

Fig. 11.4 A typical Windows start-up screen

Fig. 11.5 The AutoSketch for Windows start-up icon

The Main window

The **Main** window in the Windows screen shows icons for the various configuration facilities for the software, e.g. that labelled **Control Panel** brings up a window in which the icons represent the various Windows configuration control files. A *double-click* on the Printers icon brings the **Printers** window on screen. Using the various settings from this window allows the operator to configure the printer(s) connected to the computer; the icon labelled **Read Me** in the **Main** window brings a **Write** file on screen with notes of the

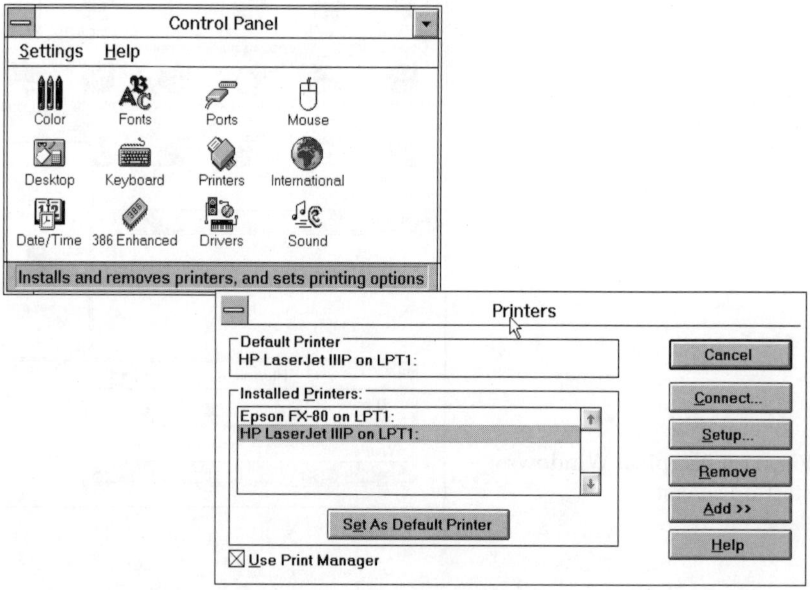

Fig. 11.6 The **Windows Control Panel**

latest details available about operating the Windows systems. The results of a *double-click* on these icons are shown in Figs. 11.6 and 11.7.

It must be remembered that Windows settings will affect AutoSketch for Windows settings. As an example, printing/plotting of AutoSketch drawing files is carried out through the Print Manager of Windows. If the printer(s) is/are not properly set, then the drawing will not be plotted/printed.

Switching between applications in Windows

One of the greatest advantages of working in Windows, especially if using Windows 3.1, is the ability to switch between applications with ease.

1. *Left-click* on the button at the very top left of the Windows screen and a pull-down menu appears – Fig. 11.8.

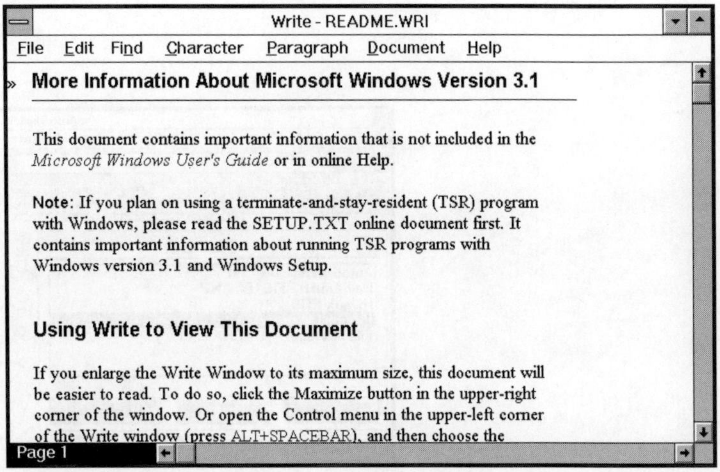

Fig. 11.7 The results of a *double-click* on **Read Me** in the **Main Window** of Windows

2. *Left-click* on **Switch To ...** in the pull-down menu which appears. A new pull-down menu appears – the **Task List** – Fig. 11.9.
3. *Left-click* on **Cascade** in the **Task List** pull-down menu: all applications in operation are cascaded on screen – Fig. 11.10.
4. *Left-click* on the button at the top right of each of the application windows in turn. The application window leaves the screen and its icon appears at the bottom of the screen – Fig. 11.11.
5. *Left-click* on each application icon in turn – the application window for the icon selected will appear back on screen.

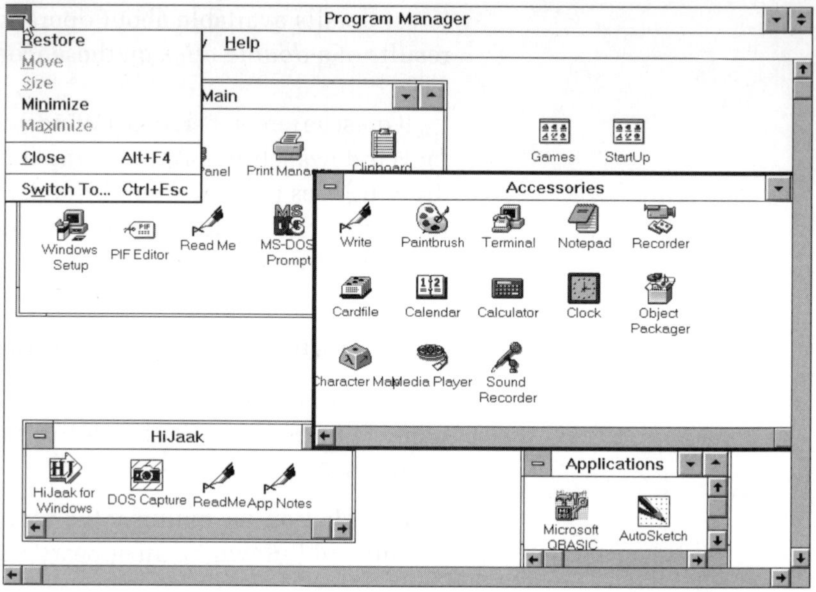

Fig. 11.8 The pull-down menu which appears after a *left-click* on the Windows top left button

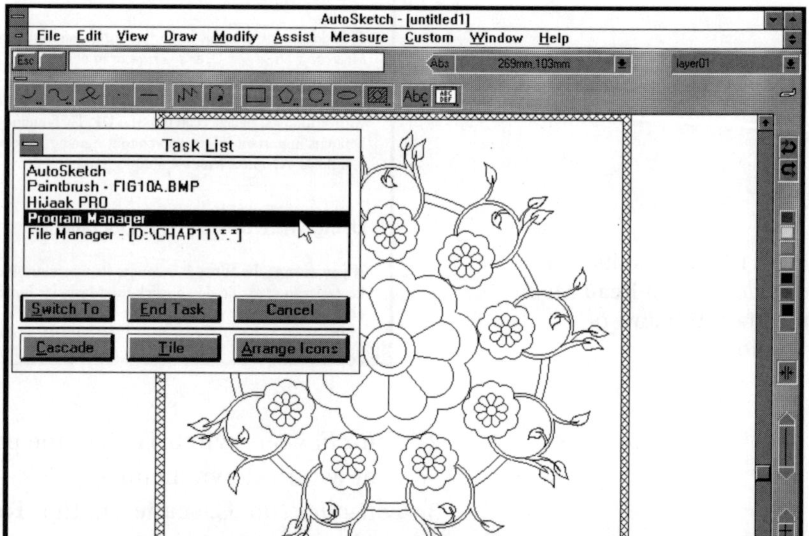

Fig. 11.9 The **Task List** pull-down menu

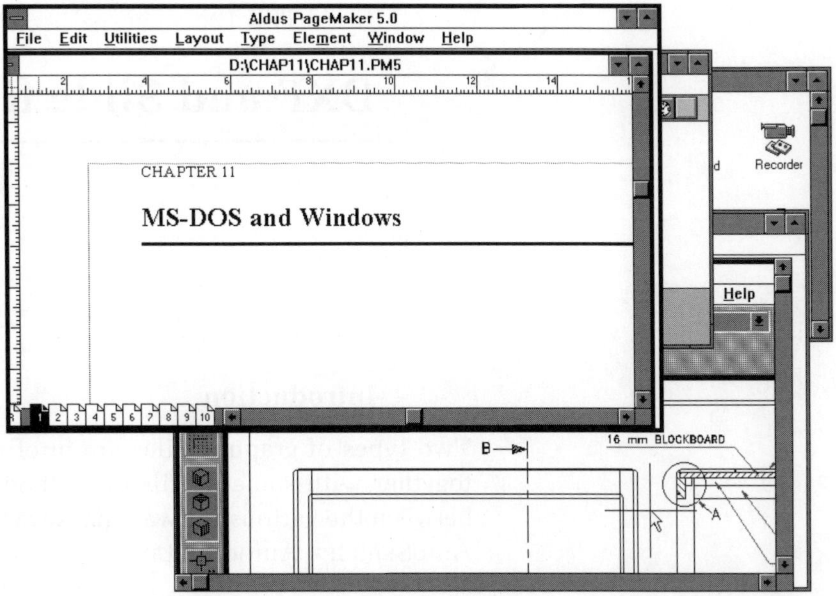

Fig. 11.10 Cascaded applications in their windows

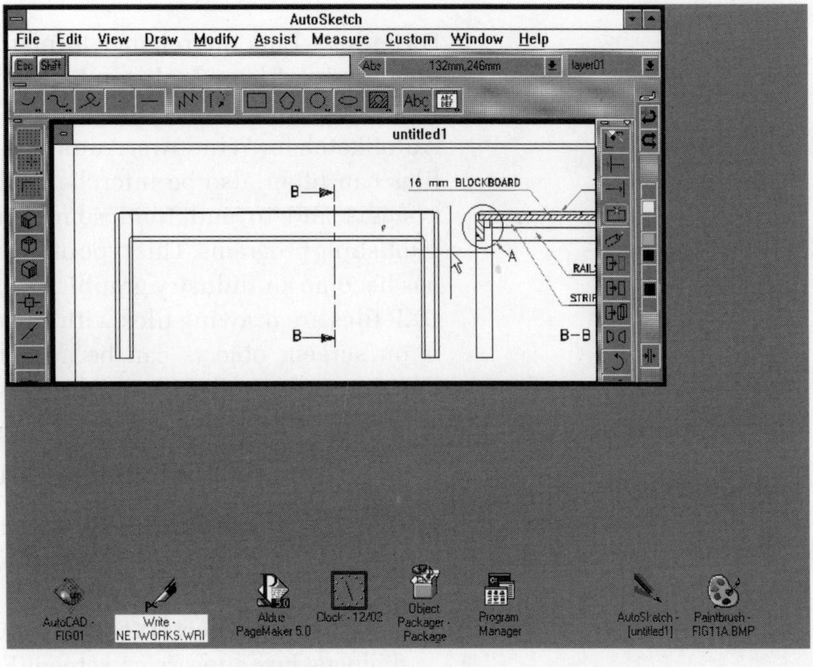

Fig. 11.11 The application icons at the bottom of the Windows window after *left-clicks* on the top right button

Note: An **application** is any item of software which, when loaded into a computer, is designed to perform certain functions, such as writing (word processors); engineering drawing and design (CAD software), desktop editors, storing data (databases), spreadsheets and the like.

CHAPTER 12

DXF and Slide files

Introduction

Two types of graphics files are briefly described in this chapter, together with some details of methods of interchanging graphics between the various software packages by the firm which produces AutoSketch – Autodesk Limited.

DXF files

DXF files – Data interchange format files (or Data eXchange Files) – are drawing files which can be interchanged between the various types of graphics software produced by Autodesk – AutoSketch, AutoSketch for Windows, AutoCAD, Generic Cadd, 3D Studio. DXF files can often also be interchanged between other CAD software systems and to and from some graphics packages and desktop publishing programs. This type of file format inaugurated by Autodesk has become an industry graphics standard interchange file system. DXF files are drawing files with an extension **.dxf**. When a DXF file is on screen, objects can be added or erased from those in the drawing and saved with the drawing.

Example 1 – exporting a DXF file

1. **Open ...** one of your AutoSketch drawing files.
2. *Left-click* on **File** in the **Menu bar**. The **File** pull-down menu appears (Fig. 12.1).
3. *Left-click* on **Export DXF ...** in the **File** menu. The **Export DXF File** dialogue box appears on screen (Fig. 12.2).
4. *Enter* a file name in the **Filename** box of the dialogue box, followed by a *left-click* on the **Save** button. The file will be saved to your file name with the extension **.DXF**.

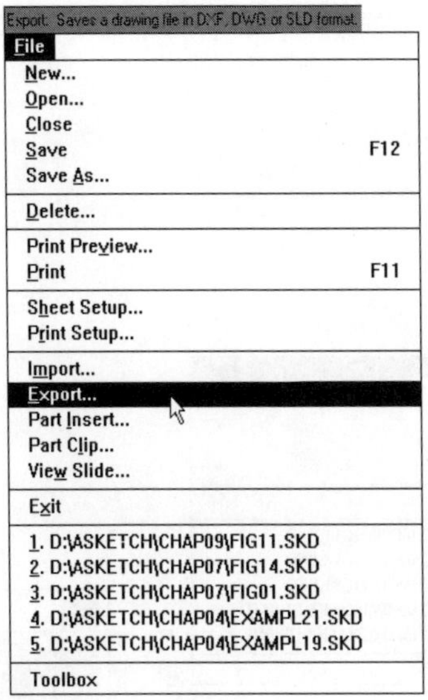

Fig. 12.1 **Export DXF ...** in the **File** pull-down menu

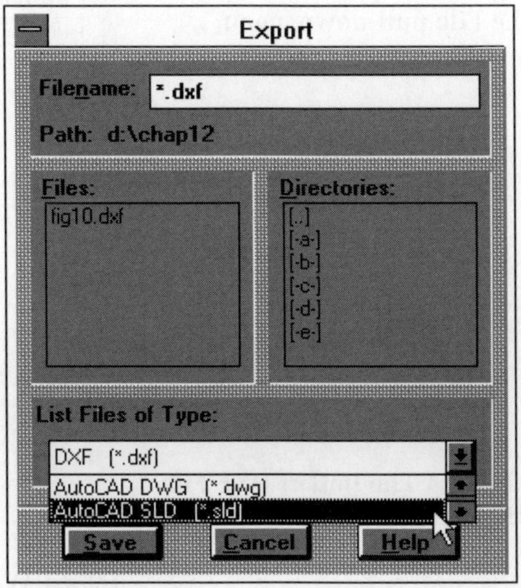

Fig. 12.2 The **Export DXF File** dialogue box

Example 2 – opening a DXF file

1. *Left-click* on **New** in the **File** pull-down menu to clear the screen.
2. *Left-click* on **File** in the **Menu bar**. The **File** pull-down menu appears (Fig. 12.3).
3. *Left-click* on **Import DXF ...** in the **File** menu. The **Import DXF File** dialogue box appears on screen (Fig. 12.4).
4. *Left-click* the name of the DXF file in the file list you have just saved, followed by a *left-click* on the **Open** button of the dialogue box.
5. The **Import DXF Control Settings** warning box appears on screen (Fig. 12.5). *Left-click* on **OK** in this box (see note below). The DXF file will appear on screen in the same way as if it were an AutoSketch drawing. There are some differences, however, as we will see below.

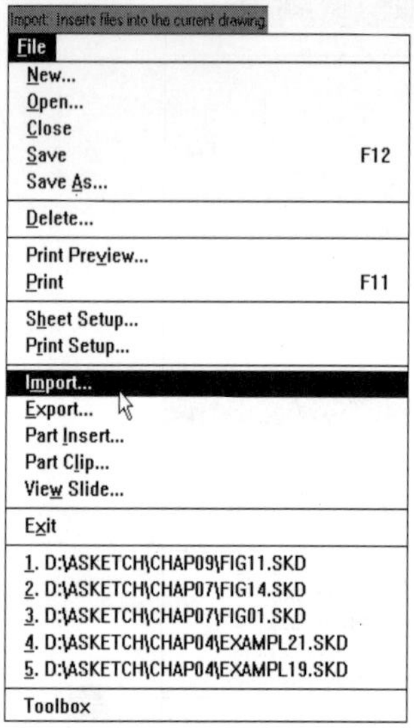

Fig. 12.3 **Import DXF ...** in the **File** pull-down menu

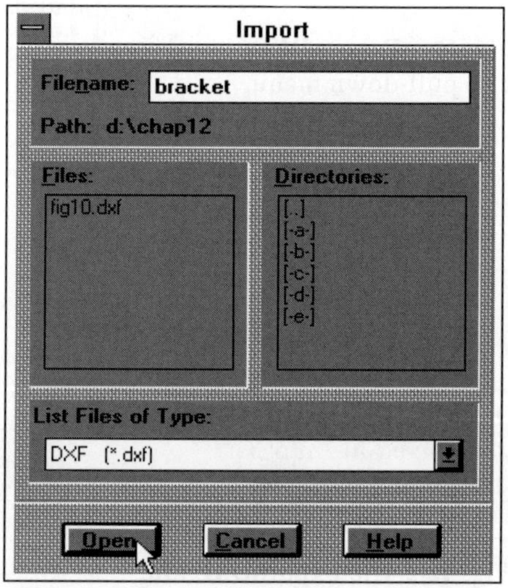

Fig. 12.4 The **Import DXF File** dialogue box

Note: We are not really concerned here with large blocks which need exploding. Neither are we concerned with Attributes. This book is really aimed at beginners and none of the recommended drawing examples or exercises included within its pages will contain large blocks or any attributes.

Examples of DXF files from AutoCAD

Example 1

Fig. 12.6 shows a simple two-view orthographic projection which has been constructed in AutoCAD Release 12. Note that the drawing contains hatching and dimensions. The drawing was saved twice as two separate DXF files. In the first saving, no attempt was made to explode anything in the drawing, although both the hatching and

Fig. 12.5 The **Import DXF Control Settings** warning box

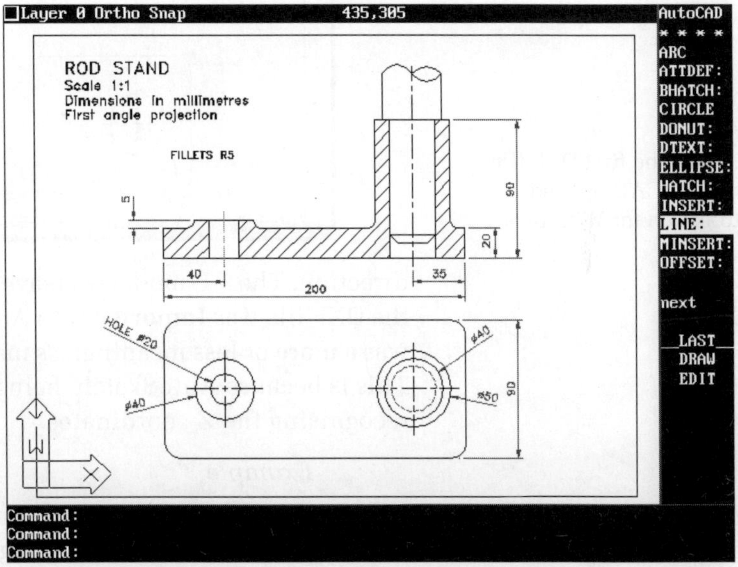

Fig. 12.6 A drawing constructed in AutoCAD Release 12

the dimensions are blocks. In the second DXF file saved in AutoCAD, each dimension was exploded in turn, i.e. the dimensions were broken by the use of the AutoCAD command **Explode** into their separate objects (entities).

Fig. 12.7 shows the AutoSketch for Windows screen when the first DXF file was **Imported**. Note the following two features:

The dimensions are missing.

The text, which in AutoCAD was in **Romand** style, has been changed to **Standard** AutoSketch style.

Fig. 12.8 shows the AutoSketch for Windows screen when the second DXF file was **Imported**. Note that the dimensions (exploded in AutoCAD) now appear.

Example 2

Figure 12.9 shows a model constructed in AutoCAD Release 12. Note that it is 3-D (three-dimensional) – it has coordinates in the *Z*

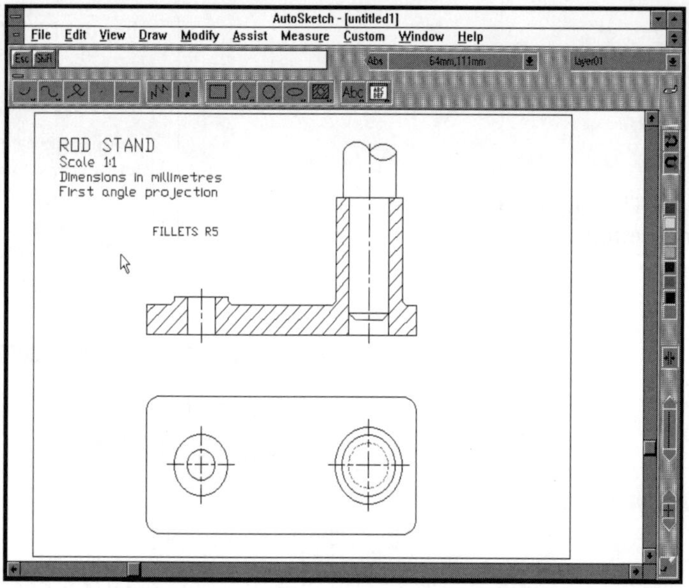

Fig. 12.7 The first DXF file from AutoCAD opened in AutoSketch for Windows

direction. The 3-D model was saved in AutoCAD as a DXF file. When the DXF file was **Imported** into AutoSketch for Windows, the result was a more or less meaningless mass of lines – as seen in Fig. 12.10. This is because AutoSketch, being a 2-D program has no facility for recognising the *Z* coordinates.

Example 3

DXF files from AutoSketch for Windows can also be loaded into 3D Studio – an excellent rendering and general graphics software

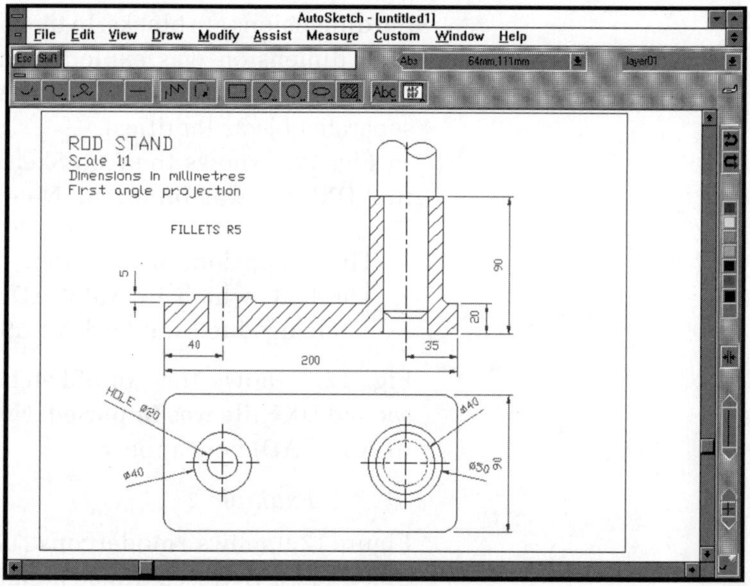

Fig 12.8 The second DXF file from AutoCAD opened in AutoSketch for Windows

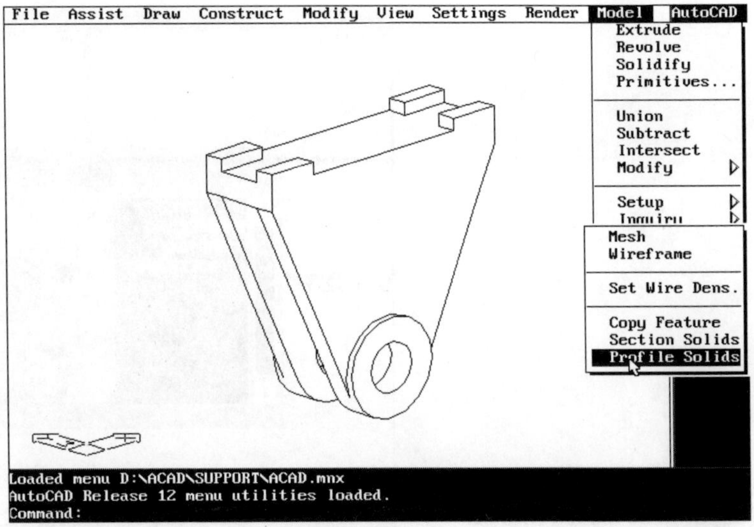

Fig. 12.9 A 3-D solid model constructed in AutoCAD Release 12

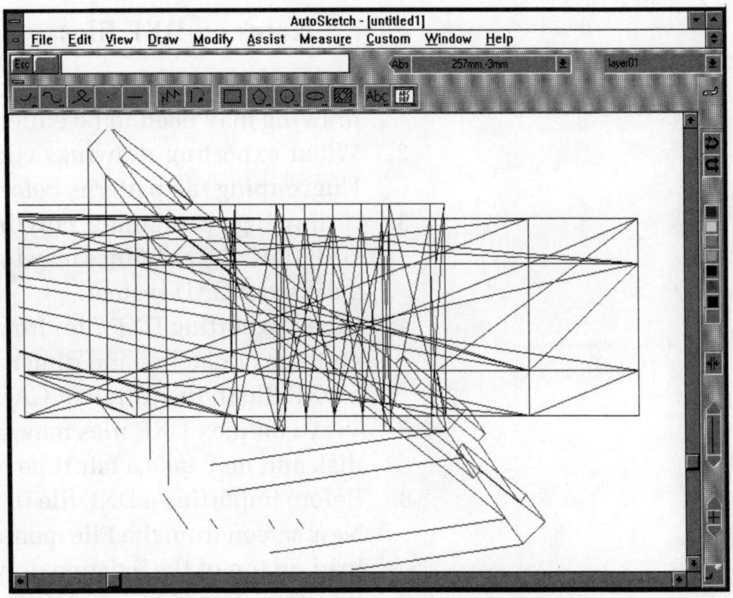

Fig. 12.10 The 3-D solid model constructed in AutoCAD, saved as a DXF file and **Imported** into AutoSketch for Windows

package published by Autodesk. These files have a limited use in 3D Studio and the software in any case is too advanced for this book, but readers may have at some time a use for importing AutoSketch DXF files into 3D Studio. Fig. 12.11 shows the **Select a mesh file to load** dialogue box from 3D Studio with some AutoSketch DXF files showing in the file list.

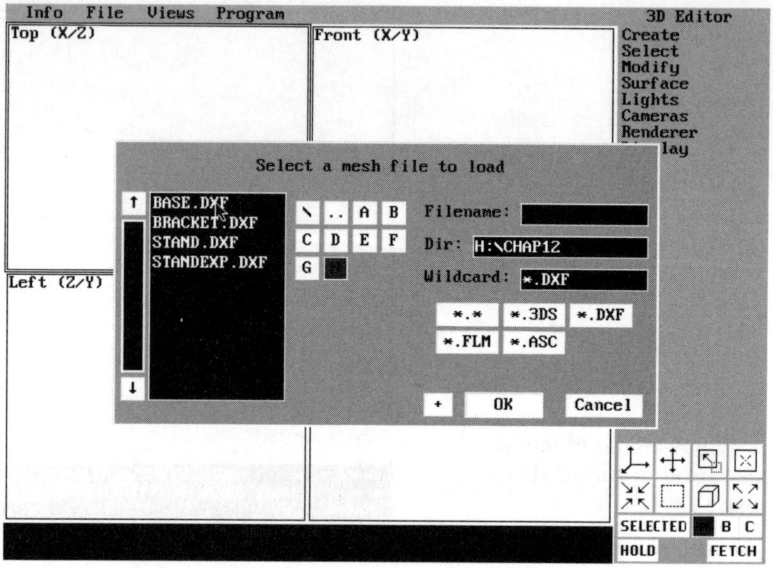

Fig. 12.11 An AutoSketch DXF file about to be loaded into 3D Studio

Notes on DXF files

1. When exporting DXF files to a different CAD system, the exported drawing may need to be edited.
2. When exporting drawings containing **Pattern Fill** patterns, try **Ungrouping** the patterns before saving as a DXF file.
3. If drawings containing **Polylines** with patterns are exported as DXF files, the polylines may become filled in the file imported to a different CAD system.
4. When exporting DXF files from a different CAD system, explode patterns or hatches and dimensions before saving the drawing as a DXF file in the different CAD system.
5. Very complex DXF files may take up a large amount of space on disk and may take a fair time to **Export** to disk.
6. Before importing a DXF file into AutoSketch for Windows, call a **New** screen from the **File** menu. Otherwise the DXF drawing will load on top of the existing drawing on screen.
7. If features in an imported DXF drawing are foreign to AutoSketch for Windows, they are ignored and will not appear on screen.
8. AutoSketch ignores all 3-D information in a DXF file, but will accept all 2-D information from a DXF file of a 3-D model from another system.
9. Centre lines, hidden detail lines and other forms of lines from AutoCAD drawings are converted into solid lines when imported via DXF into AutoSketch. To change such lines back to their original form, use the **Property (Change)** tool, with suitable line and colour settings.

10. AutoCAD layers are named in a different manner to AutoSketch layers, which are numbered 1 to 10. In exchanging DXF files from AutoCAD to AutoSketch, AutoCAD layer 0 is converted to AutoSketch layer 1; other layers in AutoCAD drawings are converted to layers from 2 to 10.

Slide files

Slide files, unlike DXF files, are not drawing files. If an attempt is made to add new objects in a slide file on screen, it appears at first as if the newly drawn objects become part of the slide. It will soon be seen, however, if the slide and its new objects are saved and then opened, that the slide part of the saved file has been lost. Objects cannot be erased from a slide file on screen. Slide files cannot be plotted or printed. Slide files are, however, easy to save and can be a useful method of showing on screen a number of drawings quickly and in the middle of constructing another drawing. If a slide is viewed while a drawing is on screen, the slide disappears with *left-click* on the **Redraw** icon. Slide files have a file extension **.SLD**.

Example 3 – creating a Slide file

1. **Open ...** any drawing.
2. *Left-click* on **Export Slide ...** in the **File** menu – the **Export Slide File** dialogue box appears on screen (Fig. 12.12).

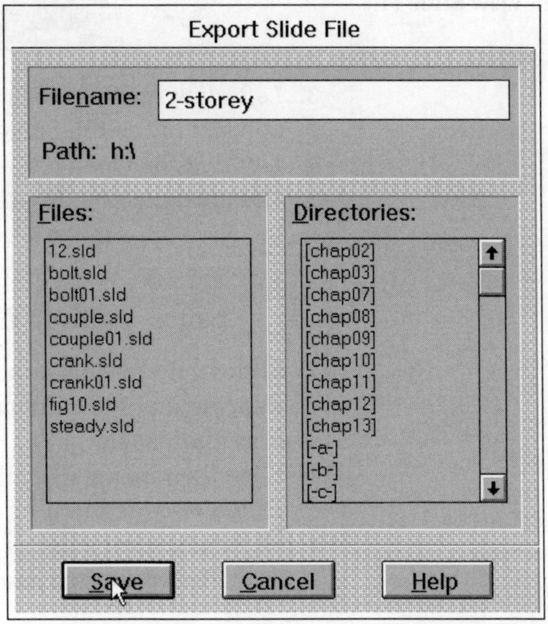

Fig. 12.12 The **Export Slide File** dialogu box

3. *Enter* an appropriate file name in the **Filename** box of the dialogue box, followed by a *left-click* on the **Save** button.
4. The file is saved with a file name extension of **.SLD**.

Example 4 – viewing a Slide file

1. **Open ...** any drawing.
2. *Left-click* on **View Slide ...** in the **File** menu – the **View Slide File** dialogue box appears on screen (Fig. 12.13).

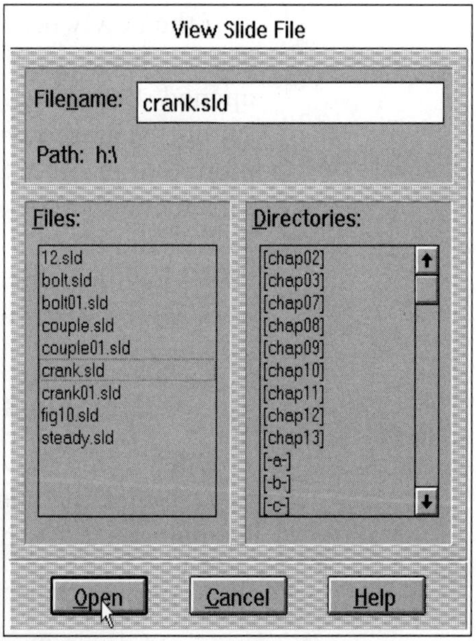

Fig. 12.13 The **View Slide File** dialogue box

3. *Left-click* on the name of the slide file to be viewed. Its name appears in the **Filename** box of the dialogue box.
4. *Left-click* on **Open** box and the slide appears covering the drawing already on screen. Alternatively, *double-click* on the file name and the slide appears on screen.

Notes on Slide files

1. Slides on screen are somewhat ephemeral. A *left-click* on the **Redraw,** any **View Zoom,** or a **Pan** will cause the slide to disappear from the screen.
2. Slide files are not like either drawing or DXF files – they cannot be edited or added to with the aid of the **Edit** or **Draw** tools.
3. Slide files form a useful 'snapshot' of drawings. For example, if a drawing is **Exported** as a slide file, it can be **Viewed** at any time for

reference purposes, without interfering in any way with the constructions taking place on screen. The viewed slide just overlays the existing drawing and can be removed by a **Redraw** leaving the original drawing intact.

4. AutoSketch for Windows slide files can be viewed in other CAD systems and in some desktop publishing programs.
5. Slides are useful for the making of presentations – a series of slides may be shown one after the other with the aid of the **Record Macro** and **Play Macro** tools of the **Utility** toolbox. A record of slides in this manner can form the basis of a good reference library, but which one can scan through one's drawings quickly, without interfering with the drawing under construction. A record of slides can also form the basis for a graphics presentation at a lecture or at a seminar.

CHAPTER 13

Arc Mode, Dim Slide and other tools

Introduction

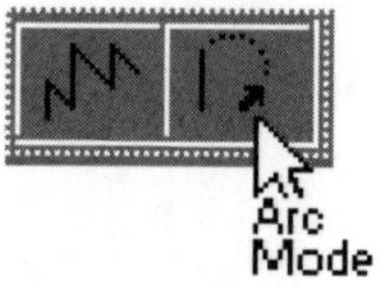

Fig. 13.1 The **Arc Mode** icon from the **Assist** toolbox

This chapter is devoted to those tools in AutoSketch for Windows which have not, as yet, been described in earlier pages. These tools are **Arc Mode**, **Dim Slide**, **Fill**, **Frame**, **Show/Hide Pop-ups**, **Help** and **Pan**.

Example 1 – Polyline and Arc Mode

1. **New ...** A3SHEET.SKT.
2. *Double-click* on the **Polyline** icon in the **Draw** toolbox.
3. *Enter* 0.7 in the **Line Width** box of the **Polyline Settings** dialogue box, followed by a *left-click* on **OK**.
4. Polyline First point: 50,200. *Left-click* on the **Arc Mode** icon in the **Assist** toolbox (Fig. 13.1) – turns tool on.

   Polyline Point on arc: 150,200
   Polyline Arc segment endpoint: 200,200

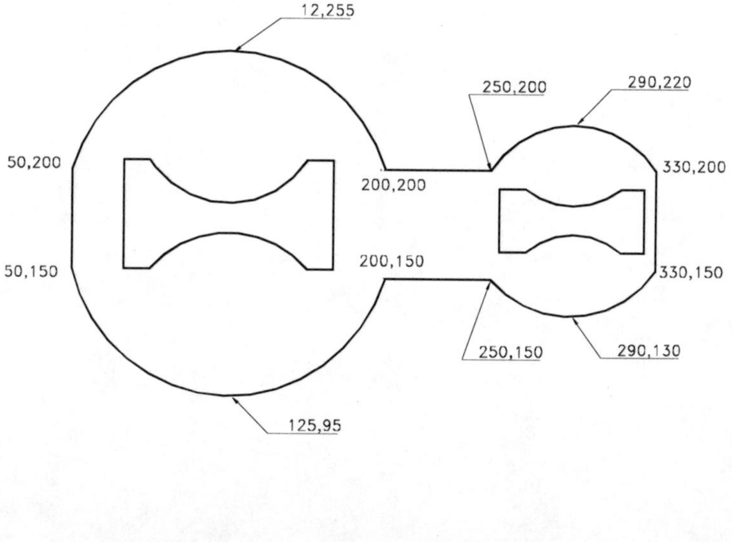

Fig. 13.2 Example 1

Left-click on the **Arc Mode** icon in the **Assist** toolbox – turns tool off.

Polyline To point: 250,200

Left-click on the **Arc Mode** icon in the **Assist** toolbox – turns tool on.

Polyline Point on arc: 290,220
Polyline Arc segment endpoint: 330,200

Left-click on the **Arc Mode** icon in the **Assist** toolbox – turns tool off.

Polyline To point: 330,150

Left-click on the **Arc Mode** icon in the **Assist** toolbox – turns tool on.

Polyline Point on arc: 290,130
Polyline Arc segment endpoint: 250,150

Continue in this manner until the outline shown in Fig. 13.2 is constructed.

Example 2 – the Fill tool

1. Construct the upper drawing of Fig. 13.3 – draw the outline with the aid of the **Polyline** tool from the **Draw** toolbox, adding the pattern with the **Pattern Fill** tool from the **Draw** toolbox. The chosen pattern for this example was **ellpaver** with a **Scale** of **2**.
2. *Left-click* on the **Fill** icon in the **View** toolbox. The pattern disappears from within its polyline boundary – lower drawing of Fig. 13.3.

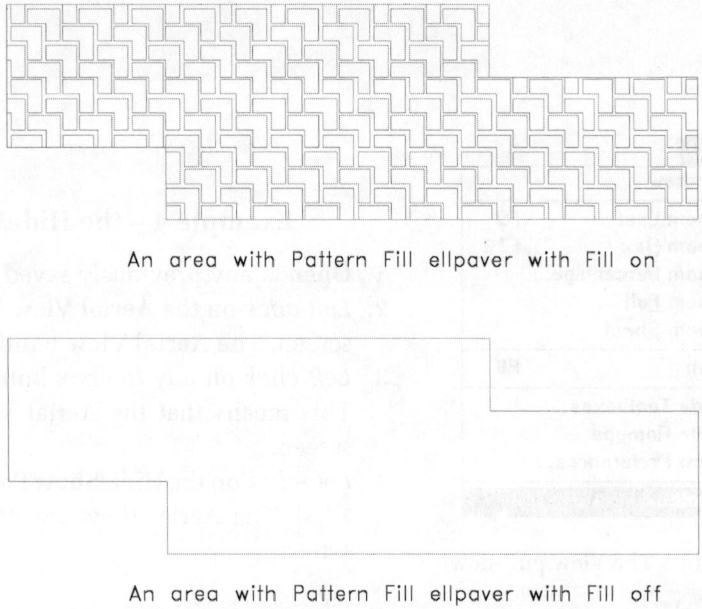

Fig. 13.3 Example 2

Example 3 – using the Frame tool

1. *Left-click* on the **Curve** icon in the **Draw** toolbox. Draw any curve with its aid.
2. *Left-click* on **Frame** in the **View** toolbox (or menu as shown in Fig. 13.4).
3. Note that a frame appears made up of lines from each of the selected points for drawing the curve.
4. With **Frame** active, draw another curve. Note that this construction will consist of only the **Frame** lines, until the curve is completed (*double-click* at end of curve). The upper two curves of Fig. 13.4 show the curves with their frames (**Frame** on), the lower two without their frames (**Frame** off).

Fig. 13.4 Example 3

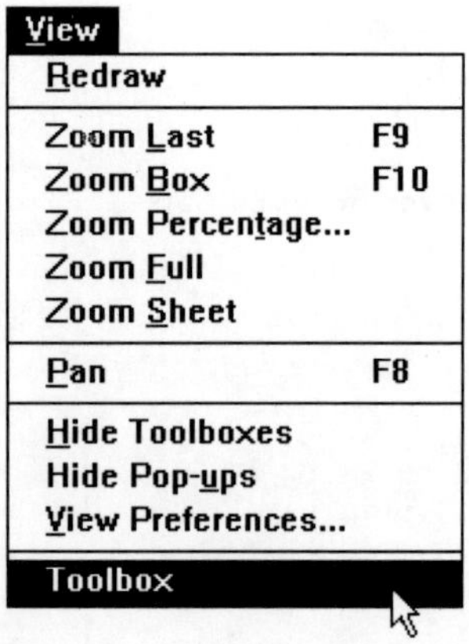

Fig. 13.5 The **View** pull-down menu

Example 4 – the Hide/Show Pop-ups tool

1. **Open ...** any previously saved drawing.
2. *Left-click* on the **Aerial View** button at the top right corner of the screen. The **Aerial View** window appears.
3. *Left-click* on any toolbox button and cause the toolbox to **Float**. This means that the Aerial View and a floating toolbox are on screen.
4. *Left-click* on the **Hide/Show Pop-ups** icon in the **View** toolbox (Fig. 13.5). The Aerial View and the floating toolbox disappear from screen.

Example 5 – the Hide/Show Toolboxes tool

1. Before starting this example, *left-click* on each toolbox button in turn and change all toolboxes into pull-down menus.
2. *Left-click* on **Hide/Show Toolboxes** in the **View** menu – or toolbox – Fig. 13.6. All toolboxes currently on screen disappear from screen.

Note that **Hide/Show Toolboxes** is a useful tool for ensuring you have the maximum size drawing area on screen, but use pull-down menus instead of toolboxes.

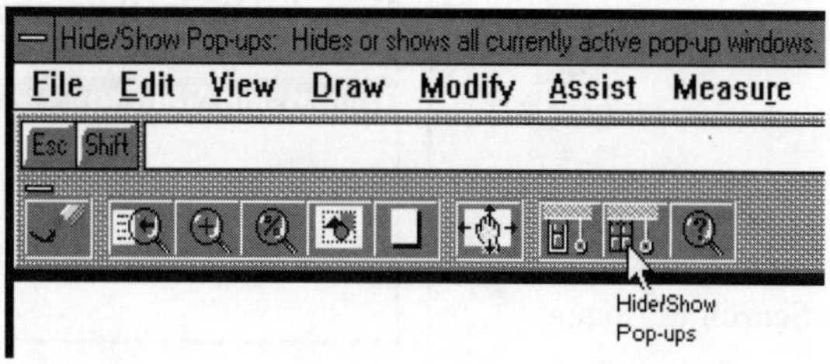

Fig. 13.6 The **Hide/Show Pop-ups** icon in the **View** toolkit

Example 6 – the Help tool

1. *Left-click* on **Help** in the **Status Bar**. The **Help** toolbox appears (Fig. 13.7).

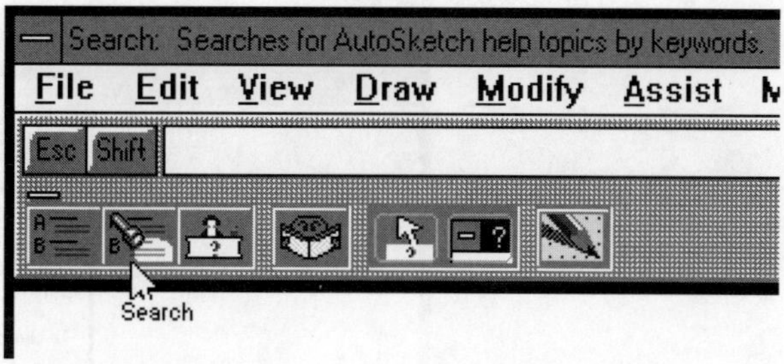

Fig. 13.7 The **Search** icon in the **Help** toolkit

2. *Left-click* on the **Search** icon in the **Help** toolbox. The **Search** dialogue box appears (Fig. 13.8).
3. *Enter* the name of the feature for which you are seeking help. In the given example **dim slide** has been entered.
4. *Double-click* on **Dimension Alignment** in the box at the bottom of the dialogue box. The **AutoSketch Help** dialogue box appears with text help for **Dimension Settings** which includes details of **Dim Slide** – Fig. 13.9.

Example 7 – the Pan tool

1. **Open ...** any previously saved drawing.
2. *Left-click* on the **Pan** icon in the **View** toolbox.

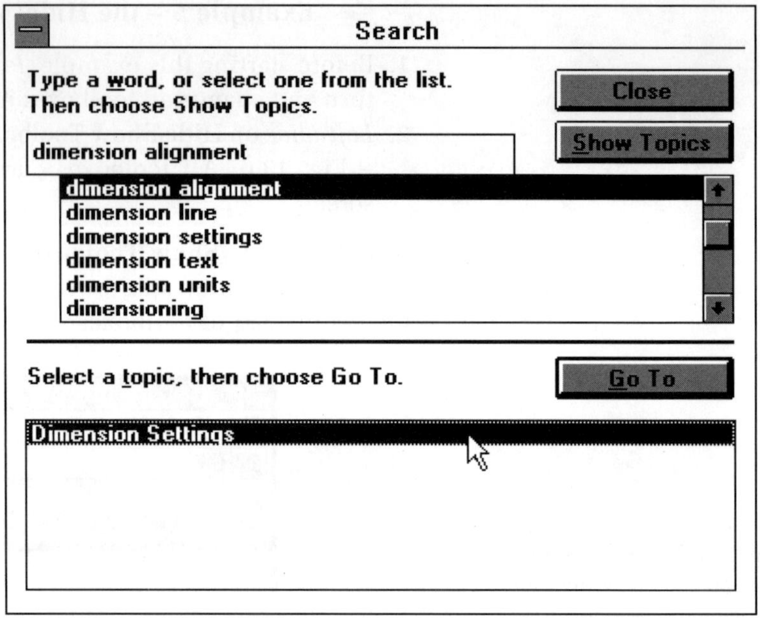

Fig. 13.8 The **Search** dialogue box

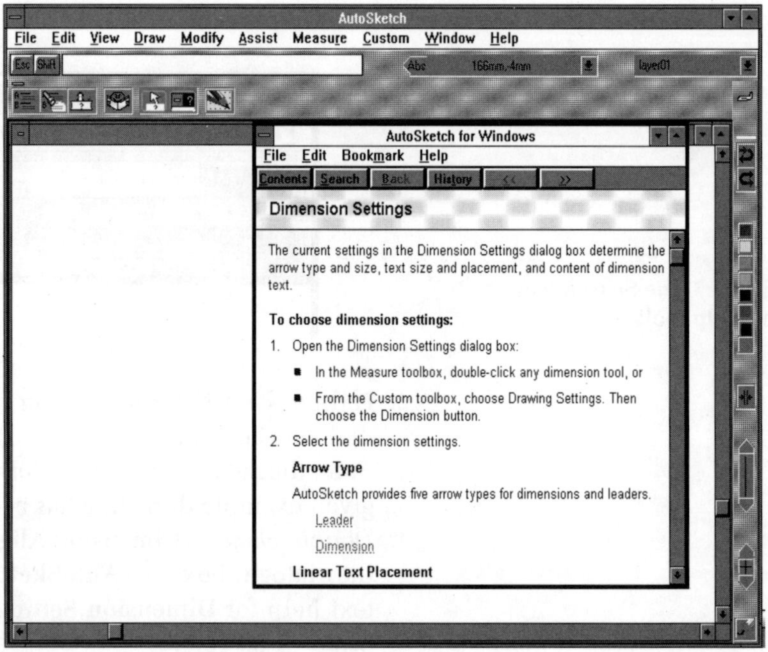

Fig. 13.9 The **AutoSketch Help** dialogue box

Pan Pan Reference: *left-click* anywhere in the drawing on screen
Pan Pan Destination: *left-click* on another point on screen and note how the whole drawing is moved between the two selected points

Note: The use of **Pan** allows you to work anywhere in a large drawing without the need for a zoom. You can always get back to the **Limits** of the drawing on screen by selecting **Sheet** from **View**.

Underlined letters in pull-down menus

You will note in Fig. 13.5, which shows the **View** menu, that all the tool names have one letter in each name underlined. Instead of a *left-click* on a tool name, the *entering* of the underlined letter at the keyboard brings the tool into operation.

Note, however, that this facility only applies when a pull-down menu is on screen. *Entering* an **F**, for example, when the menu is not on screen produces no action in AutoSketch for Windows at all.

CHAPTER 14

Printing and plotting

When you wish to print or plot your drawings, it must be remembered that we are working in Windows. This means that any printing or plotting is under the control of the Windows Print Manager. In AutoSketch for Windows, printing and plotting are both carried out under the control of the Windows Print Manager.

Setting up the printer or plotter

1. Either **Switch To...** the Windows Print Manager, or on opening Windows, *left-click* on **Main**, followed by *left-click* on **Control Panel**, followed by *left-click* on **Printers** (Fig. 14.1).
2. *Left-click* on the required printer or plotter in the **Installed Printers** box. Check that **Use Print Manager** is on (cross in its box). *Left-click* on **Cancel**. The selected printer should be on-line ready for the next stage in printing or plotting.

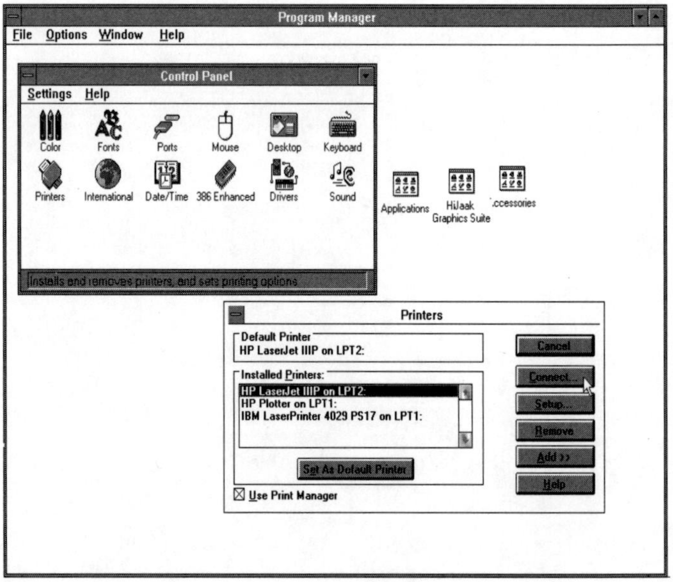

Fig. 14.1 The **Printers** windows setup to print on a laser printer

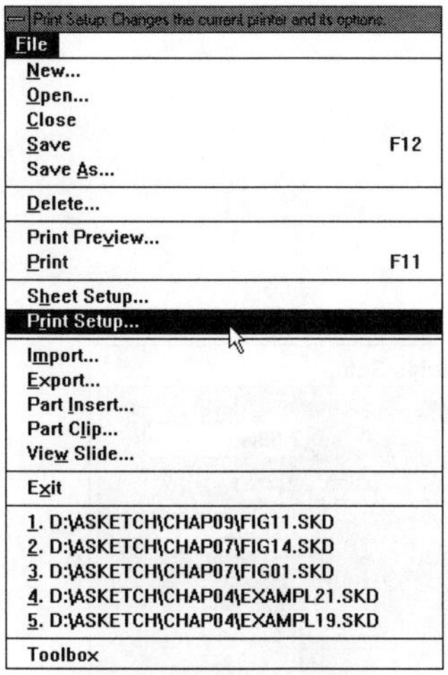

Fig. 14.2 The **File** menu showing the **Print Setup...** selection

Note: It will be seen in the **Installed Printers** box that one of the named devices is a plotter. The AutoSketch drawings in this book were all originally printer to a Hewlett Packard laser printer. However, if a plotter is available, AutoSketch drawings can equally well be plotted using the Windows **Print Manager** with AutoSketch for Windows settings.

1. **Open...** any one of your previously saved drawings.
2. *Left-click* on **Print Setup...** in **File** in the **Menu Bar** (Fig. 14.2), or *left-click* on the **Print Setup** icon in the **File** toolbox (Fig. 14.3). The **Print Setup** dialogue box appears (Fig. 14.4).
3. *Left-click* on the **Setup...** button of the dialogue box. The dialogue box for the current printer appears. *Left-click* on the **OK** button.
4. *Left-click* on the **Margins...** button of the **Print Setup** dialogue box. The **Margins Setup** dialogue box appears (Fig. 14.5). The margins can be set in this box. *Enter* the required figures in the **Left**, **Right**, **Top** and **Bottom** boxes. Examples of the results of different settings are shown later in this chapter.

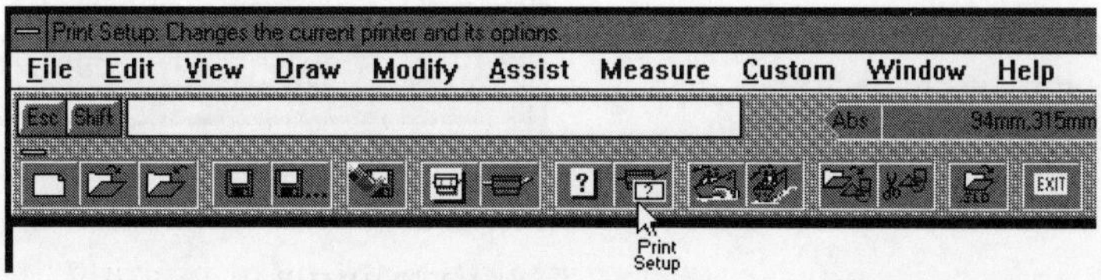

Fig. 14.3 The **Print Setup** tool icon in the **File** toolbox

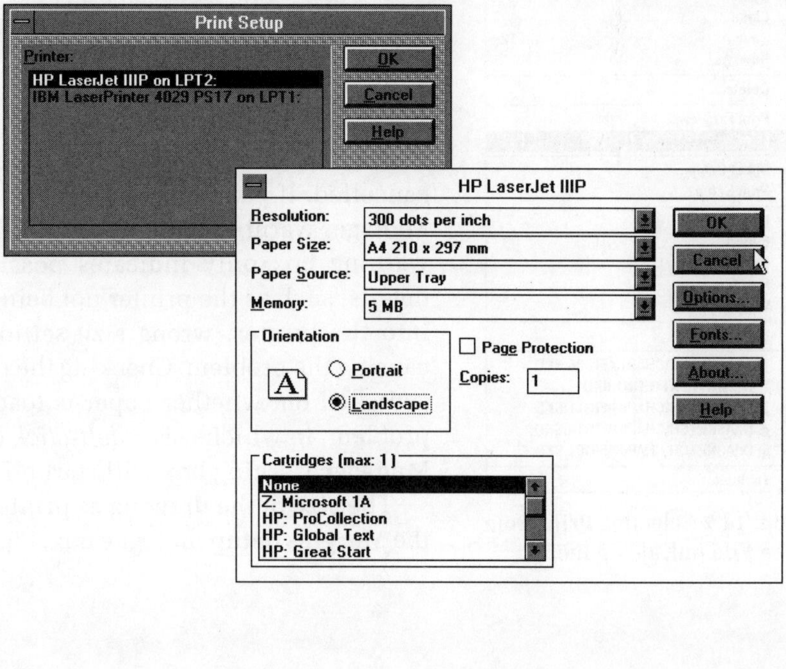

Fig. 14.4 The **Print Setup** dialogue box, together with the **Printer** details dialogue box

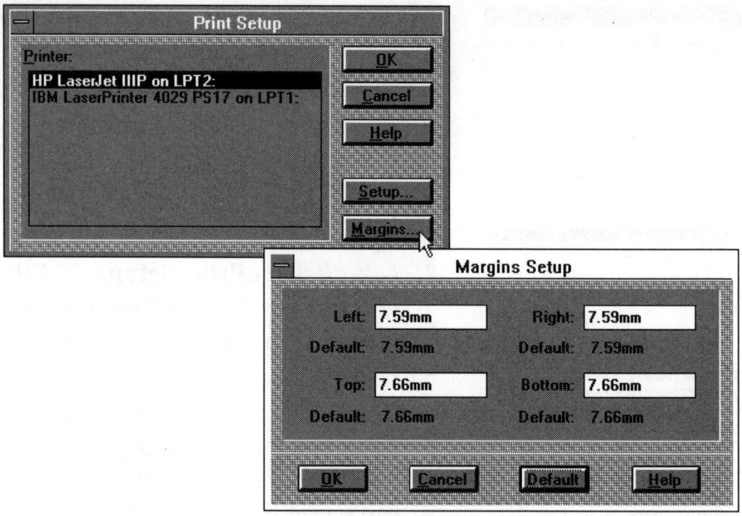

Fig 14.5 The **Print Setup** dialogue box, together with the **Margins Setup** dialogue box

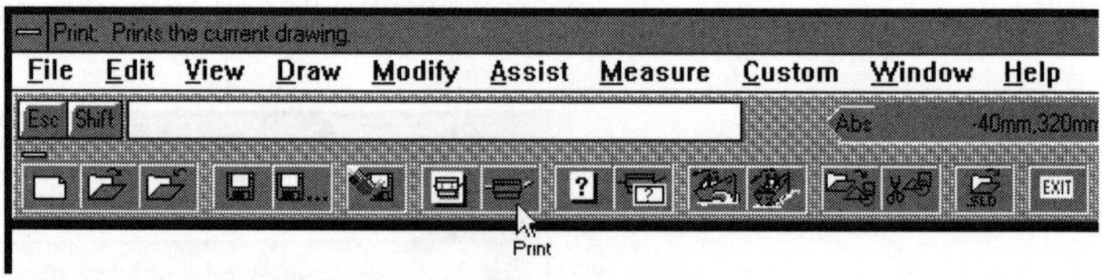

Fig. 14.6 Selecting **Print** from the **File** toolbox

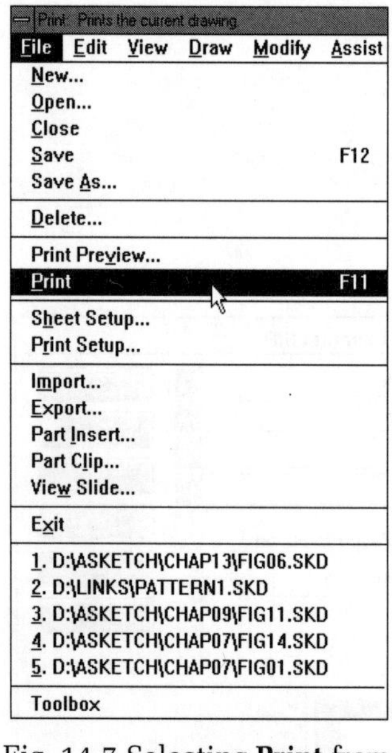

Fig. 14.7 Selecting **Print** from the **File** pull-down menu

Printing

Left-click on **Print** icon in the **File** toolbox (Fig. 14.6) or on **Print** in the **File** pull-down menu (Fig. 14.7). The **Print Preview** dialogue box appears (Fig. 14.8) showing the margins as set in the **Margins Setup** dialogue box in broken lines bounding the drawing to be printed. *Left-click* on the **OK** button of the dialogue box and the **Printing** warning box appears (Fig. 14.8). The printing now proceeds unless the **Cancel** button is selected, when the printing process will be cancelled. If, for some reason, printing cannot continue, the **Print Manager** warning box (Fig. 14.9) will come on screen. Although the warning box only indicates possible faulty port settings, other details, such as the printer not being switched on, no paper loaded into the printer, wrong size settings in the printer etc., may be causing the problem. Checking the printer, its settings, whether it is switched on, whether paper is loaded, etc., will usually clear the problem, in which case a *left-click* on the **Retry** button of the **Print Manager** warning box will start off the printing.

The size of the drawing as printed depends upon the settings in the **Margins Setup** dialogue box. Fig. 14.10 shows the **Print Preview**

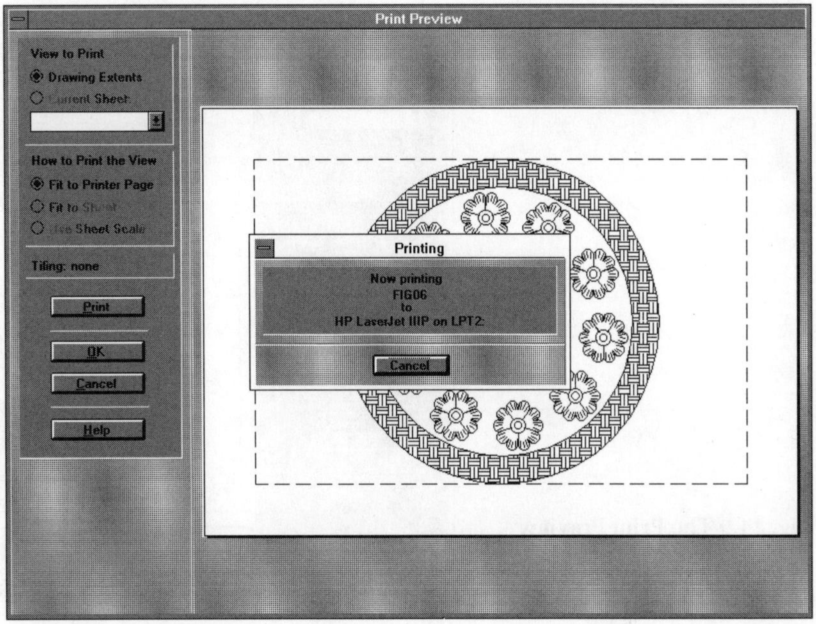

Fig. 14.8 The **Print Preview** dialogue box

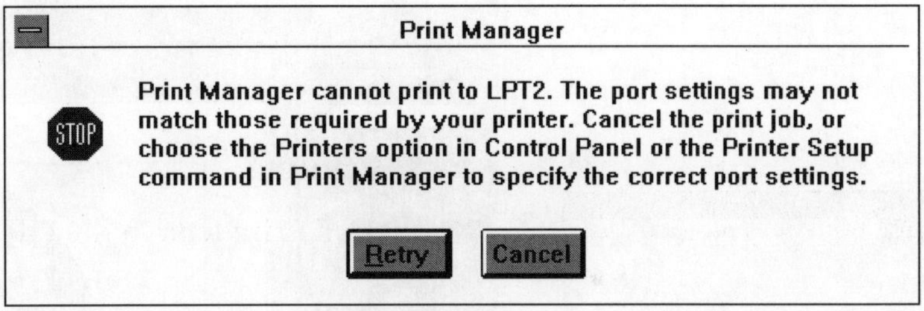

Fig. 14.9 The **Print Manager** warning box

dialogue box with each of the margins set to the default settings. Fig. 14.11 shows the **Print Preview** dialogue box with each of the margins set to 25 and Fig. 14.12 with margins of 60. The previews determine the size to which the drawing will be printed in relation to the size of the paper upon which the printing will take place.

Note: If a plotter is attached to the computer rather than a printer, the plotting will follow the same routines as for printing.

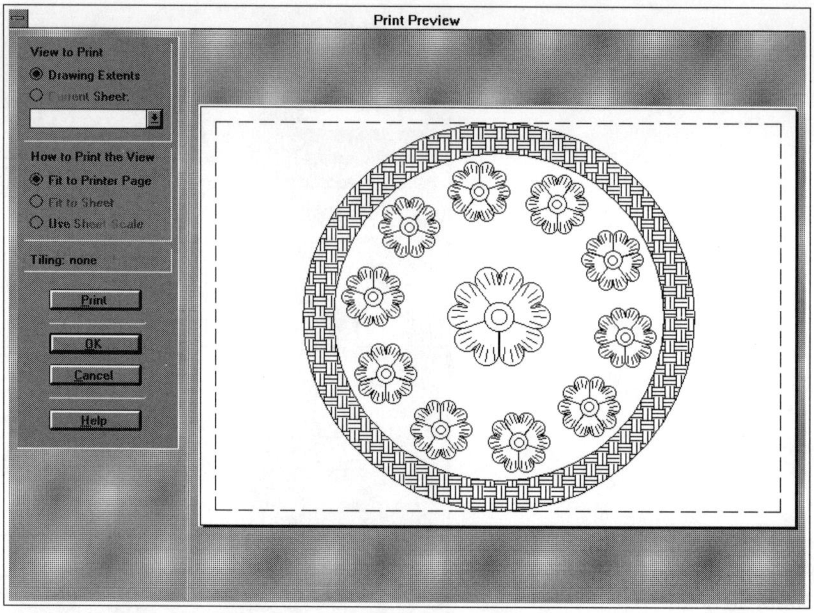

Fig. 14.9 The **Print Preview** dialogue box with margins at the default settings

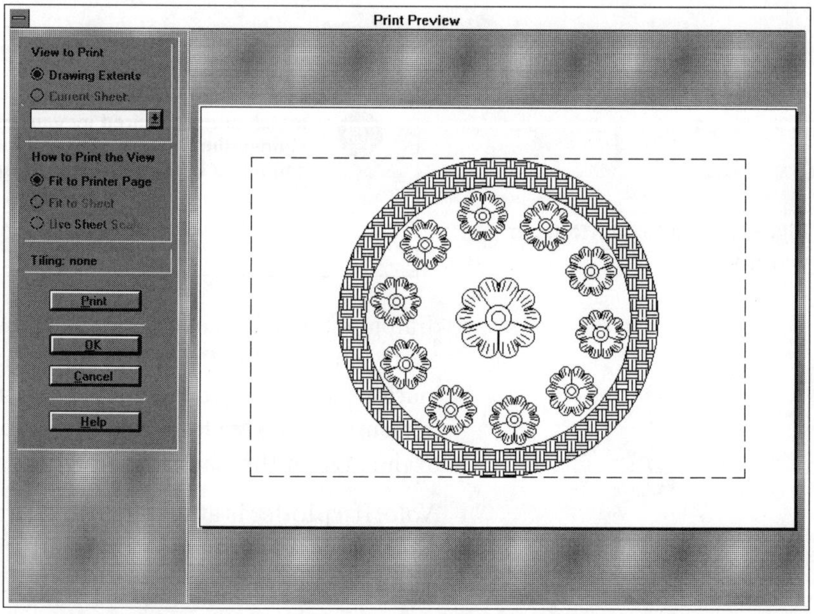

Fig. 14.11 The **Print Preview** dialogue box with margins at 25

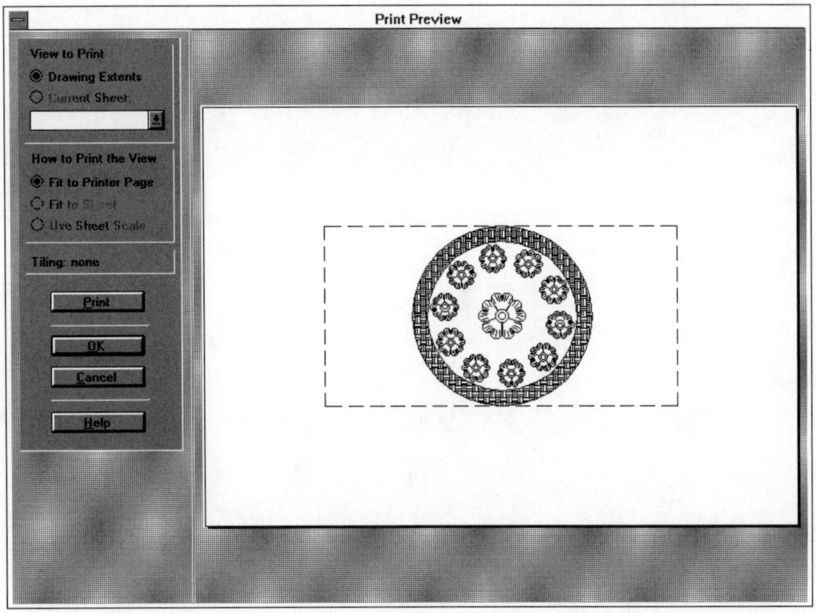

Fig. 14.12 The **Print Preview** dialogue box with margins at 60

Index